BIOMES ATLASES

OCEANS
AND BEACHES

Trevor Day

Steck-Vaughn Company

First published 2003 by Raintree Steck-Vaughn Publishers,
an imprint of Steck-Vaughn Company.
Copyright © 2003 The Brown Reference Group plc

Library of Congress Cataloging-in-Publication Data

Day, Trevor.
    Oceans and Beaches / Trevor Day.
        p. cm. -- (Biomes atlases)
    Summary: Describes the oceans of the Earth, including their effect on world climate
and the plants and animals that live in them.
    Includes bibliographical references and index.
    Contents: Biomes of the world -- Oceans of the world -- Atlantic Ocean -- Oceans and
climate -- Gulf of Mexico -- Ocean plants -- Indian Ocean -- Ocean animals --
Pacific Ocean -- People and Oceans -- Arctic Ocean -- The future.
    ISBN 0-7398-5512-3
    1. Ocean--Juvenile literature. 2. Beaches--Juvenile literature. [1. Ocean. 2. Beaches. 3.
Marine ecology. 4. Ecology.] I. Title. II. Series.

GC21.5 .D42 2003
551.46--dc21

2002068057

Printed in Singapore. Bound in the United States.
1 2 3 4 5 6 7 8 9 0 LB 07 06 05 04 03 02

The Brown Reference Group plc
Project Editor: Ben Morgan
Deputy Editor: Dr. Rob Houston
Consultant: Dr. Barbara Ransom,
    Research Scientist, Scripps Institution
    of Oceanography, University of
    California at San Diego
Designer: Reg Cox
Cartographers: Mark Walker and
    Darren Awuah
Picture Researcher: Clare Newman
Indexer: Kay Ollerenshaw
Managing Editor: Bridget Giles
Design Manager: Lynne Ross
Production: Matt Weyland

Raintree Steck-Vaughn
Editor: Walter Kossmann
Production Manager: Brian Suderski

Front cover: Tuvalu, South Pacific.
Inset: Hawksbill turtle.

Title page: Cannon Beach, Oregon.

The acknowledgments on p. 64 form
part of this copyright page.

About this Book

The introductory pages of this book describe the biomes of the world and then the ocean biome. The five main chapters look at different aspects of oceans and beaches: climate, plants, animals, people, and future. Between the chapters are detailed maps that focus on major parts of the oceans. The map pages are shown in the contents in italics, *like this*.

Throughout the book you'll also find boxed stories or fact files about oceans. The icons here show what the boxes are about. At the end of the book is a glossary, which explains what all the difficult words mean. After the glossary is a list of books and websites for further research and an index, allowing you to locate subjects anywhere in the book.

 Climate

 People

 Plants

 Future

 Animals

 Facts

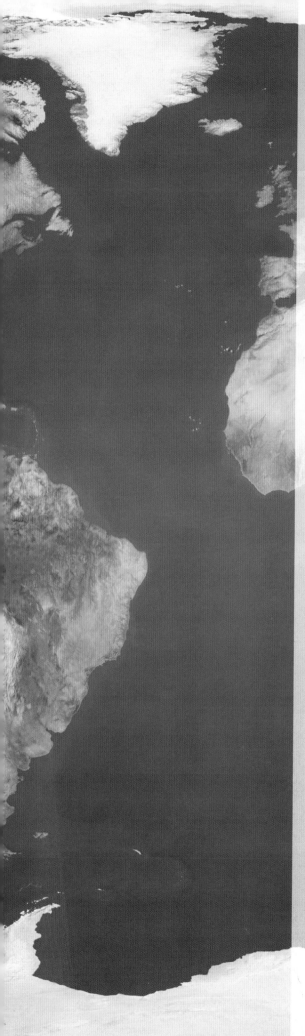

Contents

Biomes of the World

Biologists divide the living world into major zones named biomes. Each biome has its own distinctive climate, plants, and animals.

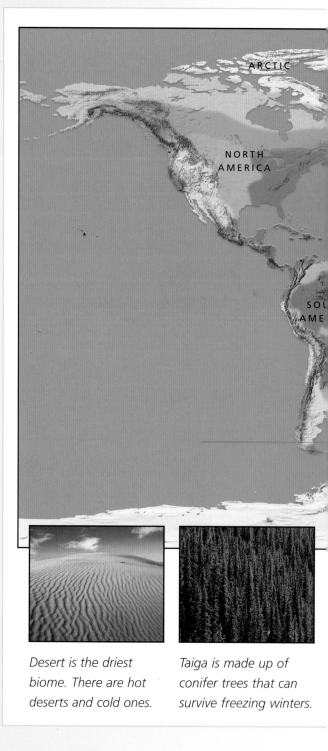

If you were to walk all the way from the north of Canada to the Amazon rain forest, you'd notice the wilderness changing dramatically along the way.

Northern Canada is a freezing and barren place without trees, where only tiny brownish-green plants can survive in the icy ground. But trudge south for long enough and you enter a magical world of conifer forests, where moose, caribou, and wolves live. After several weeks, the conifers disappear, and you reach the grass-covered prairies of the central United States. The farther south you go, the drier the land gets and the hotter the sun feels, until you find yourself hiking through a cactus-filled desert. But once you reach southern Mexico, the cacti start to disappear, and strange tropical trees begin to take their place. Here, the muggy air is filled with the calls of exotic birds and the drone of tropical insects. Finally, in Colombia you cross the Andes mountain range—whose chilly peaks remind you a little of your starting point—and descend into the dense, swampy jungles of the Amazon rain forest.

Desert is the driest biome. There are hot deserts and cold ones.

Taiga is made up of conifer trees that can survive freezing winters.

Scientists have a special name for the different regions—such as desert, tropical rain forest, and prairie—that you'd pass through on such a journey. They call them biomes. Everywhere on Earth can be classified as being in one biome or another, and the same biome often appears in lots of

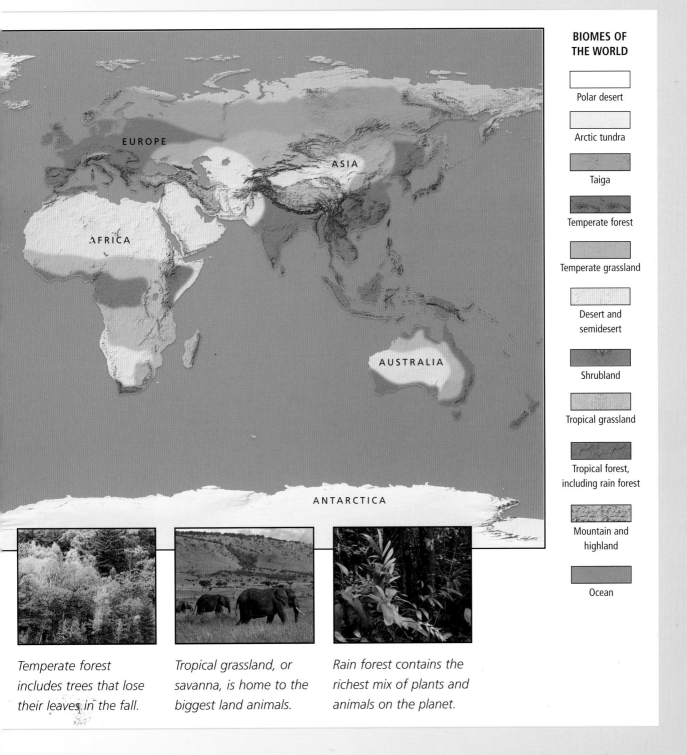

BIOMES OF THE WORLD

Polar desert

Arctic tundra

Taiga

Temperate forest

Temperate grassland

Desert and semidesert

Shrubland

Tropical grassland

Tropical forest, including rain forest

Mountain and highland

Ocean

Temperate forest includes trees that lose their leaves in the fall.

Tropical grassland, or savanna, is home to the biggest land animals.

Rain forest contains the richest mix of plants and animals on the planet.

different places. For instance, there are areas of rain forest as far apart as Brazil, Africa, and Southeast Asia. Although the plants and animals that inhabit these forests are different, they live in similar ways. Likewise, the prairies of North America are part of the grassland biome, which also occurs in China, Australia, and Argentina. Wherever there are grasslands, there are grazing animals that feed on the grass, as well as large carnivores that hunt and kill the grazers.

The map on this page shows how the world's major biomes fit together to make up the biosphere—the zone of life on Earth.

Oceans of the World

Oceans cover more than two-thirds of Earth's surface, making the ocean biome the biggest biome of all. Under the waves, though, there are a multitude of habitats, from the beach to the deep-ocean floor.

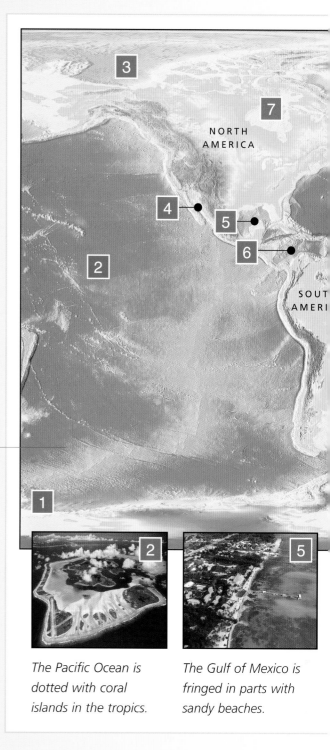

The oceans are one gigantic body of water, but we give different names to different parts. The four major oceans are the Pacific Ocean, Atlantic Ocean, Indian Ocean, and Arctic Ocean. Each lies in a huge hollow in Earth's crust, called a basin.

The oceans are not really one habitat, but a collection of habitats, many of which overlap. Most ocean plants and animals live near the coast, where the water is rich in nutrients from rivers, and where the shallow seafloor gets plenty of sunlight. Coastal habitats include forests of giant seaweeds, meadows of sea grass, and some coral reefs.

Where sea meets land, another habitat exists: the intertidal zone. It can be a violent place, where savage waves continually crash into rocky cliffs. In other places, there are calm, sandy beaches that slope gently into the water. The animals and plants of the intertidal zone lead double lives, having to cope with being exposed to air and then being covered with water, at least once a day.

The Pacific Ocean is dotted with coral islands in the tropics.

The Gulf of Mexico is fringed in parts with sandy beaches.

Most of the open ocean has far less wildlife than the coast. Even so, the sea surface in places teems with microscopic creatures called plankton, which flourish in the sunlit water. Deep below the plankton is a zone of dark, cold water less rich in life but with some very unusual inhabitants.

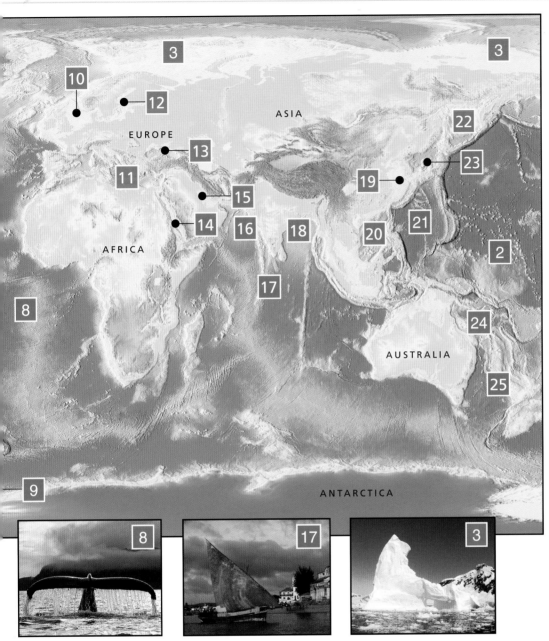

Humpback whales feed in the Atlantic Ocean, the Arctic, and the Antarctic.

Monsoon winds carry sailboats along trade routes across the Indian Ocean.

The Arctic Ocean is wide and deep, but it lies beneath a covering of ice.

The ocean floor is home to all sorts of animals. Many feed on dead matter raining from the waters above. Most seem to live near land, or on the summits of underwater mountains. Most of the seafloor has not been explored, so there could be many creatures far from land that we don't know about.

You might think the deepest parts of the ocean would be devoid of life, but they are home to some of the weirdest animals of all. While most life on Earth depends on the sun for energy, some of the strange creatures living on the deep-sea floor get their energy from chemicals gushing from hot-water vents.

Atlantic Ocean

Named after the Greek god Atlas, who held up the heavens, the Atlantic is the second largest ocean. It stretches from the Arctic to Antarctica and contains about one-quarter of the world's seawater.

 ## Atlantic Facts

▲ The Atlantic Ocean is vast—it is about four times the size of North America.

▲ The average depth of the Atlantic is a bit more than 2 miles (3.3 km), or about 20 times the height of the Washington Monument in Washington, D.C.

▲ The Atlantic is deepest at the Puerto Rico Trench. It is 5.2 miles (8.4 km) deep there, or 22 times the height of New York's Empire State Building.

▲ The Atlantic is getting wider at the rate of about 1 inch (2.5 cm) a year—about the same rate that your fingernails grow. When Christopher Columbus made his transatlantic voyage in 1492, the Atlantic was 40 feet (12 m) narrower.

 ## Atlantis Uncovered

During an expedition in the late 19th century, British scientists traveled widely across the oceans to sample their depths. The scientists discovered that parts of the Atlantic were much shallower than expected. At the time, some British newspapers declared that the scientists had found the fabled submerged city of Atlantis. In fact, they had stumbled across what was to be called the Mid-Atlantic Ridge. The east and west parts of the ocean floor are moving apart. The ridge forms where molten volcanic rock wells up from beneath Earth's crust and flows into the resulting gap. As the volcanic rock piles up, it forms a ridge. In places, the ridge is so high that it reaches the ocean surface, forming islands such as the Azores and Iceland. Today, we know that the Mid-Atlantic Ridge is just part of the huge midocean ridge system that extends around the globe.

In addition to molten rock, superheated water gushes up from the midocean ridge. The water is rich in minerals, which billow in dark clouds called black smokers (right). These smokers provide energy for strange forms of deep-ocean life.

Arctic Ocean

Svalbard

GREENLAND

N

Denmark Strait

ICELAND

Surtsey **1**

Hudson Bay

CANADA

Cod fishing grounds **2**

NORTH AMERICA

Grand Banks

● Wreck of the *Titanic*

EUROPE

Gulf Stream

UNITED STATES

3 Bay of Fundy

New England Seamounts

4 Blake Ridge

7

5

Horseshoe Seamounts

6 Madeira

Azores

Canary Islands

BERMUDA

Sargasso Sea **8**

Gulf of Mexico

BAHAMAS

CUBA

Caribbean Sea

Puerto Rico Trench

Hispaniola

Midocean ridge

Atlantic

Tropic Seamount

AFRICA

CAPE VERDE ISLANDS

Mouths of the Amazon River

SOUTH AMERICA

Ocean

Ascension

Midocean ridge

9

Columbia Seamount

Rio de Janeiro

● São Paulo

Untreated sewage **10**

Tristan da Cunha **12**

Walvis Ridge

11

Benguela Current

0 ____ 1,500 miles
0 ____ 1,500 km

Pacific Ocean

ANTARCTICA

7. Bermuda
Northerly coral reefs flourish here, warmed by the Gulf Stream.

8. Sargasso Sea
An area of ocean that remains relatively still on the surface, and where floating seaweed grows. The seaweed shelters and feeds animals, including camouflaged sargassum fish.

9. Midocean Ridge
The Atlantic is spreading outward from the midocean ridge, where new seafloor is forming. Hydrothermal vent habitats develop here.

10. Untreated Sewage
Hot spots for sewage pollution lie close to the large Brazilian cities of Rio de Janeiro and São Paulo.

11. Benguela Current
Cold, nutrient-rich water wells up here from the deep ocean. Plankton multiply and feed huge shoals of fish. Millions of sharks, tuna, and seabirds thrive on the small fish.

12. Tristan da Cunha
The most remote inhabited land on Earth. The people here live 1,500 miles (2,400 km) from the nearest human settlement, which is in South Africa.

1. Surtsey
In 1963, the volcanic island of Surtsey rose above the sea from close to the midocean ridge near Iceland.

2. Cod Fishing Grounds
In the early 1990s, the great cod fisheries of the northwest Atlantic closed because there were too few fish left to catch.

3. Bay of Fundy
The tidal range here is an enormous 52 feet (16 m)—the highest tidal range in the world.

4. Blake Ridge
A huge deposit of methane lies under the seabed here. Seeps of this gas support life-forms that obtain energy from methane. There are mussel beds the size of a football field and mussels as big as footballs.

5. The Gulf Stream
A fast, warm current that flows from the Gulf of Mexico and crosses the North Atlantic. It helps make northwest Europe 18°F (10°C) warmer than places at the same latitude in Canada.

6. The Azores
These remote volcanic islands are surrounded by seas rich with life. Twenty-five types of whales and dolphins come here to dine.

Humpback whales feed in both arctic and antarctic waters in the Atlantic Ocean.

Oceans and Climate

Oceans have a huge influence on Earth's weather. They also have their own underwater climates, which depend on ocean currents, depth below the water surface, and the way sunlight travels through water.

Seawater absorbs sunlight. Even in the clearest water, only about 1 percent of the light reaches a depth of 330 feet (100 m), and only the very surface of the ocean receives as much light as land. Sunlight is a mixture of different colors, which together appear white. The sea absorbs the different colors in differing amounts, screening some out but letting others pass. If you go scuba diving in a tropical sea, you soon notice that the spectacular coral reef colors get less vivid the deeper you go. Red and yellow get filtered out, making everything look blue-green.

Beyond the Twilight Zone

Scientists divide the ocean into layers, or zones, based on depth. The top zone is the sunlit zone. As you sink through it, the light fades, the water gets cooler, and the pressure increases. At a depth of 660 feet (200 m), there is barely any light left, and the pressure is 20 times higher than at the surface. This is the bottom of the sunlit zone.

Conditions on the coast are dominated by the tides and the force of the waves. Over time, waves can carve bare rock into towering sculptures called sea stacks, like these at Cannon Beach in Oregon.

Beneath the sunlit zone, down to 3,300 feet (1,000 m), lies the twilight zone. The creatures of the twilight zone are often red or black, making them almost invisible in the gloom. But there are flashes of light because many animals, including twilight-zone fish and squid, produce their own light by a process called bioluminescence. They use this light to dazzle predators, to attract prey, or to signal to other members of their species.

Beyond 3,300 feet (1,000 m) deep is the dark zone—a world of total blackness, except for occasional flashes of bioluminescent light. This zone extends to the deep plains that make up most of the ocean floor. Pressures in the dark zone are more than 100 times greater than at the surface—enough to crush your lungs completely—and the water is bitterly cold, hovering between freezing point and 39°F (4°C). Some of the animals here are

Ocean Facts

▲ Oceans cover 71 percent of Earth's surface.

▲ The seas and oceans are slowly changing shape and size. The Pacific Ocean is shrinking, but the Red Sea is getting wider.

▲ If all the continents were scraped off to sea level and dumped into the Pacific basin, they wouldn't come close to filling it in.

Ocean Zones

Gannets

Right: Air-breathing divers usually stay less than 160 feet (50 m) from the surface. Even at these depths, seawater filters out most red light, leaving only blue and green.

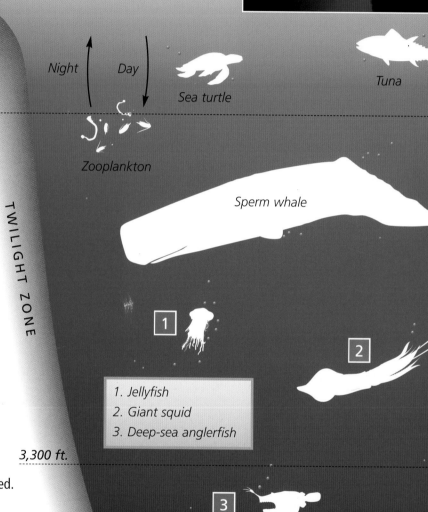

TIDAL ZONE

Peacock worm

Crab

Reef-building coral

Clam

BEACH

OPEN OCEAN

SUNLIT ZONE

Night   Day

660 ft.

Sea turtle

Tuna

Zooplankton

Sperm whale

TWILIGHT ZONE

1. Jellyfish
2. Giant squid
3. Deep-sea anglerfish

1

2

3,300 ft.

3

DARK ZONE

Deep-sea brittle star

Deep-sea sponge

Deep-sea lil

Tidal Zone

Whether it is rocky, sandy, or muddy, the zone between high tide and low tide is home to creatures that attach themselves, burrow into the sediment, or scuttle on top.

Sunlit Zone

On the seabed below low tide lie coral reefs or sea-grass beds in the tropics, or kelp forests in cooler seas. Tiny plantlike plankton drift in the surface waters, attracting grazers, which in turn attract predators such as tuna.

Twilight Zone

Many open-ocean animals from tiny zooplankton to massive sharks hide in these dark depths during the day but visit the surface at night to feed.

Dark Zone

Mysterious creatures wander these great depths looking for occasional morsels. These are hunting grounds for squid and anglerfish.

Seabed

There is life on the seabed at all depths. Even on the deep-ocean floor there is a thin rain of organic matter falling from the zones above that is filtered by sea lilies and sponges or scavenged by brittle stars.

Oceans play a central role in Earth's climate. Water escapes from the ocean surface as water vapor. It rises, cools, and turns to droplets, which form clouds.

into deep trenches, reaching depths of 6 miles (10 km) or more. Life in the abyssal zone is similar to that in the dark zone, but pressures here are even more extreme. Only a handful of people in the deepest-diving submersibles have ever entered this world.

Heat Sink

Compared to land, oceans warm up and cool down very slowly. As a result, they act as a gigantic heat store, able to absorb vast amounts of energy from the sun in warm weather, and then release the heat slowly during cold weather. The oceans therefore have a huge influence on Earth's climate.

In warm, sunny parts of the world, the oceans pick up heat from the sun. Ocean currents then carry this heat away and release it in cooler areas. Warm seawater flowing toward the North Pole, for example, keeps the Arctic Ocean warmer than it would otherwise be. Besides carrying heat around

miniature monsters: grotesque fish with swollen heads, gaping mouths, and scrawny bodies. Food is so scarce that these fish try to eat everything they come across, dead or alive. It is an eat-or-be-eaten world.

There is another zone below the dark zone: the abyssal zone. This is the deepest part of the ocean, where the seafloor plunges

Origins of Life

Life on Earth probably began in the sea around 4 billion years ago. There are several reasons why scientists think this. First, organisms are at least 70 percent water, and the concentration of salt and other chemicals in their body liquids is similar to that of seawater. Second, the oldest fossils so far discovered are stromatolites— stony formations built up by mats of sea-living bacteria 3.8 billion years ago. Bacteria still build stromatolites in certain parts of the ocean today (right).

The first life might have arisen in shallow coastal waters warmed by the sun, where stromatolites grow. However, some scientists suspect that life may have begun on the deep-ocean floor, powered not by the sun but by chemical energy.

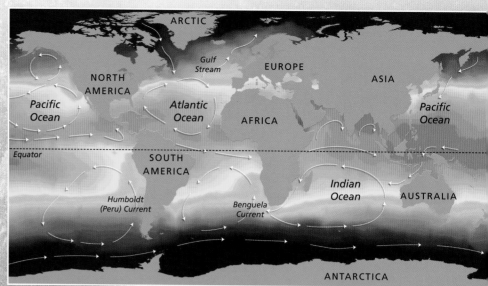

Right: Sea surface temperatures range from below 40°F (4°C), shown in black on this map, to above 80°F (27°C), shown in red. The warmest waters are near the equator, but where there are strong currents (white arrows), the ocean can be warmer or cooler than expected. The cold Humboldt (Peru) Current off the west coast of South America, for instance, causes a tongue of cool water to extend hundreds of miles toward the equator.

ⓘ Rogue Waves

In the days of sail, it was vital that seafarers knew the courses of major ocean currents. Sailing against a current could add many days to a journey, and sailing across a current could deflect a ship from its course. Even today, seafarers have to understand ocean currents. If waves from a storm meet an oncoming current, the colliding waves can add together to form "rogue waves" big enough to swamp ships. Such rogue waves, some of which sailors claim are more than 100 feet (30 m) high, sometimes form off the southeast coast of Africa—a notorious danger spot where even giant tankers have been known to sink.

the planet, the oceans release water vapor into the air, producing the clouds and rain on which life on land depends.

Rivers in the Sea

There are several reasons why ocean currents exist. One is simply that winds blowing on the ocean surface push the water along.

Winds exist because Earth's air is in constant motion, driven by the sun's energy and stirred around by our planet's rotation. Air is continually circulating, traveling up and down in great circular patterns called circulation cells. The bottoms of these cells, where air moves along the planet's surface, form global winds. These, in turn, produce global currents in the ocean surface. One such current is the Gulf Stream, which flows across the Atlantic from the Gulf of Mexico to western Europe. The Gulf Stream transports nearly 500 million bathtubs full of water every second—several times the water passing through all the world's rivers put together. All that water carries a lot of heat, which keeps the climate of northwest Europe much milder than it would be otherwise.

Merry-Go-Round

The major ocean currents flow in huge circles called gyres (shown on the map opposite), which are set spinning by Earth's rotation. Earth's rotation also makes the gyres turn in specific directions. In the northern hemisphere, the gyres turn clockwise. In the southern hemisphere, they turn counterclockwise. One consequence of

El Niño

In most years, strong winds blow from east to west across the middle of the Pacific Ocean, driving the ocean current west. As warm surface water moves west, cool water wells up from deep below to take its place, bringing nutrients to the surface off the coast of Peru. The upwelling supports such vast shoals of anchovies that it forms one of the world's most important fishing sites. But in some years (left), the winds are weak. The upwelling does not develop, the anchovies don't come, and the fishing is disastrous. The cool, nutrient-rich water is replaced by warm water from the western Pacific (white and red in the image). This is called an El Niño event. The El Niño event alters wind patterns around the world, causing unusual and extreme weather. It also makes the sea surface warmer in some places, killing the organisms that build coral reefs.

NORTH AMERICA

cool water

warm water

PERU

Pacific Ocean

this is that in the southern Atlantic and Pacific, cold water from the poles extends along west coasts, while warm water from the equator usually travels along east coasts.

Upwelling and Downwelling

Life in the oceans is richest where there is plenty of light, warmth, and nutrients. The microscopic plants that live in surface waters—phytoplankton—need these to grow. But many parts of the open sea are like deserts because of a lack of nutrients. Where there are few nutrients, there are few phytoplankton. And without phytoplankton, there are few of the animals that eat them.

There are oases in this desert. In certain places, a current of cold water from deep below, a process called upwelling, brings nutrients to the surface. Plankton grow in abundance at these places, providing food for fish and other animals. People also benefit from regions of upwelling—many of the world's fish are caught in these places.

Upwelling tends to happen on the eastern side of the great oceans. They are caused by offshore winds, which blow the surface waters out to sea. Deep waters full of nutrients well up from below to replace the surface waters.

What goes up must come down. Where ocean surface currents from opposite directions collide, some of the water is forced downward in a process called downwelling. The downwelling water is usually clear because it has few nutrients and not very much plankton. The clear waters let plenty of light through, which is ideal for coral reefs. Some of the world's largest coral reef systems—those of the Caribbean and the Great Barrier Reef of Australia, for example—grow in regions of downwelling on the western side of oceans.

Oceanic Conveyor Belt

Downwelling and upwelling areas are linked by deep currents flowing sluggishly through the oceans many thousands of feet down. These currents form what scientists call the oceanic conveyor belt—a continuous current of water that carries heat energy around the planet. The conveyor belt carries huge volumes of water and probably has a great effect on Earth's climate.

One downwelling region that feeds the oceanic conveyor belt is where the Atlantic Ocean meets the Arctic. Surface currents carry warm water from the Atlantic to the Arctic, where the water cools, freezes, and floats. Salt is left behind when seawater freezes, so the remaining water is cold and salty, which makes it heavy. This cold water sinks and begins its long voyage southward along the oceanic conveyor belt.

Between the Tides

A rocky shore can be packed with all kinds of life, from seaweeds to crabs and sea anemones (right). A beach seems like a good place to live. But compared to the many thousands of different organisms that live in shallow water close to the shore, fewer species thrive on the beach. Why? Because life on a beach is tough.

A rocky beach is a frontier between land and sea. When the tide is out, the beach is exposed to the air—which can be far hotter or colder than the seawater. Marine organisms can dry out. If it is raining, animals are drenched in freshwater rather than salt water. And when beach animals are exposed to air, they can be attacked by land predators.

Life is still tough when the tide returns. Animals are battered by waves, and marine predators can then attack them. A barnacle or mussel fixed to a rock

cannot run away when conditions change—it has to stay there and put up with the changes. Sea anemones are also stranded. They choose hollows that become rock pools at low tide, or attach themselves in crevices that stay moist, withdrawing their tentacles to protect them from drying out. For those who can ride it out, the sea brings a fresh supply of food with each rising tide, and for seaweeds there is plenty of sunlight.

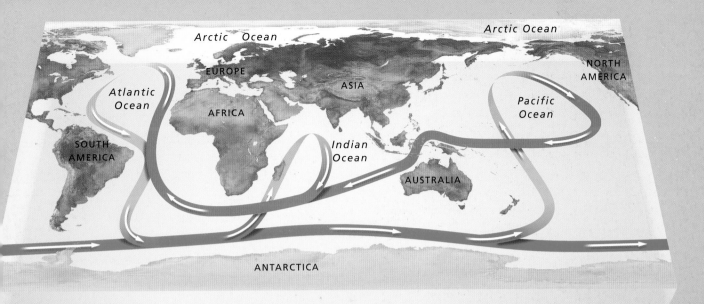

Warm surface currents (red) and cold, deep-sea currents (blue) are connected in a continuous stream of moving water called the oceanic conveyor belt.

Global warming could disrupt this process. If ice stops forming in the far North Atlantic, the oceanic conveyor belt would be interrupted, with consequent climatic effects occurring across the globe.

Hot or Cold

Temperature is one of the major factors that affects where sea creatures live. Birds and mammals are warm-blooded—they keep their bodies at a constant warm temperature, no matter whether the surroundings are warm or cool. Warm-blooded animals such as whales can travel between warm and cold areas and still survive. The gray whale, for instance, spends summers in the chilly waters of the Arctic, but swims to much warmer waters off Baja California in winter. Most sea animals, however, are cold-blooded, which means their body temperature is greatly affected by their surroundings. Such species are less flexible about where they can live. Tropical fish would die in cold water, for instance, and cold-water fish cannot survive in tropical water.

The Bay of Fundy, Nova Scotia, Canada, has the largest tidal range in the world. During the maximum high tide, the water is 52 feet (16 m) higher than at low tide.

Gulf of Mexico

From the sandy beaches and coral reefs of the coast to the mysterious chemical-energy communities of the deep-ocean floor, the Gulf of Mexico provides varied habitats for many kinds of marine life.

 Fact File

▲ Almost the entire world population of Kemp's ridley turtles—the rarest sea turtle in the world—arrives each year at Rancho Nuevo beach in Mexico. The turtles' purpose is to lay eggs together.

▲ Divers discovered the phenomenon of mass coral spawning in the Gulf of Mexico. Once a year, all the coral polyps release their eggs and sperm at once, creating an upside-down snowstorm. Scientists now study mass coral spawning the world over.

1. Rancho Nuevo
A beach where 95 percent of Earth's population of Kemp's ridley turtles come to nest.

2. Flower Garden Banks
This U.S. National Marine Sanctuary protects an area of coral reefs. In 1990, divers here observed the snowstorm effect of a mass coral spawning for the first time.

3. Methane Seep
Scientists found ice worms, in 1997, living in a methane seep, where methane leaks into the ocean from under the seabed.

4. Meso-American Barrier Reef System
The second largest barrier reef on Earth lies off the coast of Belize. Countless billions of coral polyps built the reef by laying down their stony skeletons as they grew.

5. Yucatán Channel
Water flows into the Gulf of Mexico here, carrying plankton from the coral reefs of Belize.

6. Florida Keys
A long chain of low islands with coral reefs on the south side and sand and sea-grass beds on the north. Currents from the Yucatán Channel bring the young of tropical life-forms from the Caribbean Sea and deposit them on the coral reefs of the Florida Keys.

7. Florida Bay
The sea-grass beds here are among the largest on Earth. Some of them are threatened by pollution. The beds are grazed by slow-moving manatees (sea cows).

8. Florida Strait
Water flows out of the Gulf of Mexico through the Florida Strait, forming a current that eventually flows to Europe.

9. Bahamas
Scientists take part in the Wild Dolphin Research Program in the warm, shallow waters around the Bahamas island chain. They observe the social behavior of the dolphins.

Palm-fringed beaches with beautifully clear water are common in the Florida Keys. The main keys are connected by the Overseas Highway, one of the longest overwater roads in the world.

NORTH AMERICA
EUROPE
ASIA
AFRICA
SOUTH AMERICA
AUSTRALIA
ANTARCTICA

0 | 300 miles
0 | 300 | km

UNITED STATES

Appalachian Mountains

● Washington, D.C.

N

Atlantic Ocean

● Houston

● New Orleans
← *Mississippi Delta*

MEXICO

FLORIDA

● Monterrey ● Rancho Nuevo
1

2 ← *Flower Garden Banks*

3 ← *Methane Seep*

● Miami

Florida Bay
7 →

Gulf of Mexico

● Mexico City
● Guadalajara

Yucatán Peninsula

5

6 → *Florida Keys*
Florida Strait

● Havana

Nassau ●

B A H A M A S

8

9

C U B A

● Belmopan
BELIZE

4 ← *Meso-American Barrier Reef System*

Yucatán Channel

CAYMAN ISLANDS
Cayman Trench ▼

HAITI

JAMAICA
● Kingston

Caribbean Sea

GUATEMALA
● Guatemala City
● San Salvador
EL SALVADOR

HONDURAS
● Tegucigalpa

NICARAGUA
● Managua

Pacific Ocean

COSTA RICA
● San José

PANAMA
● Panama City

● Barranquilla

COLOMBIA

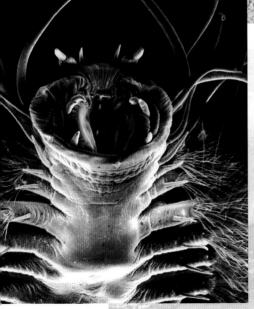

Lair of the Ice Worm

On the deep-sea floor in the Gulf of Mexico live some of the strangest animals known to science—ice worms (left). These pinkish worms, only 1–2 inches (2–5 cm) long, live in methane seeps, places where methane leaks into the ocean from deposits under the seabed. On land, methane is a gas, but in the cold and the high pressure of the seabed, it behaves strangely; it combines with water to form a kind of methane ice. It is in this unique habitat that ice worms live. They burrow in the methane ice and sculpt it into peaks like whipped cream. Scientists suspect that the worms live on bacteria. Some bacteria can use methane as a source of chemical energy to make food, just as plants use the energy in sunlight.

Ocean Plants

Most ocean plants are not true plants at all. They are algae—simple, plantlike organisms that lack proper stems, leaves, or roots. But like the true plants on land, they trap sunlight and use it to make food.

Coccolithophores are microscopic algae with intricate chalky shells made of overlapping plates. After they die, the shells fall to the seafloor, where they build up and gradually turn into chalk or limestone. These colored microscope images show coccolithophores at 2,000 times their actual size.

Plants on land are obvious—just take a look at the trees and grasses around you. But where are the plants in the oceans?

On some beaches you can find seaweeds—usually visible as brown, leathery ribbons washed back and forth by the waves. In some shallow waters, you can spot meadows of sea grass. But out at sea, you can't see any plants at all. They are there, but they are microscopic. They are single-celled marine algae. In a cupful of seawater there are tens of millions of them.

Over the course of a year, the growth of these microscopic marine algae is greater than that of all the grasses and trees on land. Marine algae are really important—it's just that most of them are too small to see without using a microscope.

Plankton

Most marine algae are plankton—organisms that drift along with ocean currents. Those plankton, including algae, that can trap sunlight and use it to make food are called phytoplankton (*phyto* means "plant").

When phytoplankton make food they also release oxygen, as do land plants. The whole process, called photosynthesis, is vital to most life-forms, because they rely on the food and oxygen that plants produce. Zooplankton (animal-like plankton) eat phytoplankton, and larger animals such as fish and squid eat the zooplankton. The vast majority of marine animals' food comes from phytoplankton.

The smallest phytoplankton are incredibly tiny—hundreds would fit on a pinhead. Many are not algae but a kind of bacteria, although they are only distantly related to the bacteria that cause disease in people. Because of their small size, these bacterial plankton slipped through scientists' nets, so nobody realized how common and important they were until recently. We now know that the tiniest

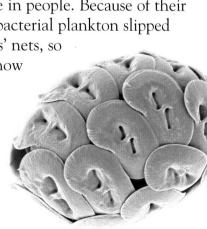

SCOTLAND

Plankton Blooms

Every spring, there is an explosion in the number of phytoplankton in the surface of the North Atlantic, brought on by warm, sunny weather. At its height, this "plankton bloom" is so intense that the plankton are clearly visible from space. The bright blue-green patch of plankton in the picture lies off the coasts of England and France. The bloom fades in summer, and the dead plankton sink, forming a sludge that settles on the ocean floor.

All manner of worms, crabs, and sea cucumbers feed on the fresh sludge. It also attracts swimming scavengers, from fish to small, shrimplike creatures called amphipods.

IRELAND

ENGLAND

Plankton bloom

FRANCE

21

Diatoms are minute algae that multiply in all the world's oceans. Their many complex shapes are formed by glassy skeletons.

phytoplankton can produce more than three-quarters of the food and oxygen in any particular part of the sea.

Small but Beautiful

The medium and large phytoplankton are tiny, too. The largest are about the size of the period at the end of this sentence. Some, called coccolithophores, look like a tiny ball covered in round, chalky plates. When these plankton die and fall to the seafloor, their remains build up over millions of years to form a chalky carpet that can be hundreds of feet thick. England's famous White Cliffs of Dover formed this way from countless billions of coccolithophores.

Many phytoplankton have fascinating shapes. Diatoms, for instance, have beautifully intricate skeletons made of silica (the main chemical in glass). Others, called dinoflagellates, have thinner skeletons shaped like leaves or cones. Many dinoflagellates are bioluminescent—they produce their own light when the water is disturbed. In some parts of the world, the sea surface sparkles at night with the green light of dinoflagellates when waves break or when a person swims.

Red Tides

Sometimes phytoplankton grow so quickly, and in such huge numbers, that they color the water. As they die and become trapped on the surface, they form a scum. This is called a red tide. Despite the name, it can be any color from yellow to blue.

Red tides can remove so much oxygen from the water that sea animals suffocate. And some red tides are poisonous. Shellfish, such as clams and oysters, can feed on the poisonous phytoplankton unharmed, but fish or other animals that eat the shellfish may die. Worldwide, red tides caused by a phytoplankton called *Alexandrium* kill nearly 200 people every year this way.

Left: This is not a river of pollution but a red tide— a mass of tiny dinoflagellates. Under certain conditions, these tiny algae multiply so quickly that they stain the sea and can poison creatures that feed on them.

Kelp forests are full of dark corners for sea life to hide, so they are good places for divers to visit. This diver is filming in a California kelp forest.

Seaweeds

You can often find seaweeds attached to rocks in cool, shallow water. There are three different types—red, green, and brown—depending on the colored chemical used to trap light in photosynthesis. The biggest seaweeds are of the brown type and are called giant kelp. They are like underwater trees, towering up from the seafloor to the surface with fronds as long as 330 feet (100 m). The smallest seaweeds, such as the bright green sea lettuce, are paper-thin. Other seaweeds form a tufted or crusty covering on rocks.

Large seaweeds have a tough time. Growing in shallow water, they have to resist the battering of waves and strong currents. They have to stay anchored to stop being swept out to sea. At the same time, they have to reach toward the surface to trap sunlight.

Below: There are many types of kelp, but they all live in cool seas. They grow nearer the equator only where cold currents bring water from cooler regions. This happens on the west coasts of South America and Africa.

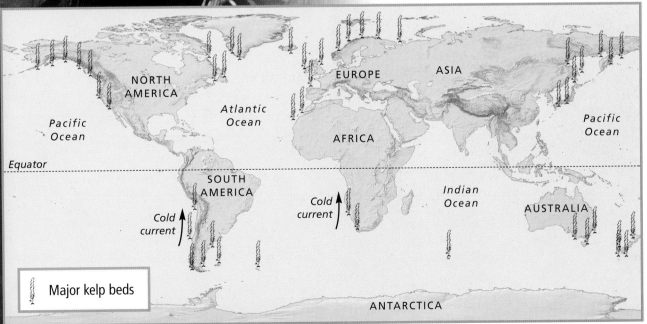

NORTH AMERICA
EUROPE
ASIA
Atlantic Ocean
Pacific Ocean
Pacific Ocean
AFRICA
Equator
SOUTH AMERICA
Cold current
Cold current
Indian Ocean
AUSTRALIA
Major kelp beds
ANTARCTICA

A large seaweed is mostly three parts: a stem, fronds or blades (like leaves), and a holdfast. The stem and fronds are flexible so they can bend with the currents and waves. If they were rigid, they would easily snap. The holdfast, which looks a bit like a tangle of roots, anchors the seaweed to rocks on the seafloor. The fronds and the stem often contain air sacs, or bladders, that make the seaweed's fronds float.

Kelp and Otters

Lots of animals live among seaweed or attached to it, although few animals eat live seaweed. Many more feed on the thin covering of green algae that grows on the seaweed's surface.

In the North Pacific, there is a strange relationship between kelp and sea otters. Where sea otters live among the kelp, the kelp usually thrives, unless the water is polluted. But where sea otters are absent (perhaps removed by hunting), the kelp seems to go through a boom-and-bust cycle. Sometimes it is abundant, but at other times

Above: After being laid out to dry, this seaweed harvested in California might end up being used as a strengthener in ceramics or as thickener in jelly.

Below: In parts of California, sea otter numbers are slowly recovering and healthy kelp forests are returning. The sea otter is so important to the kelp habitat that it is called a keystone species.

it almost disappears. Sea urchins—spine-covered animals related to starfish—seem to be responsible. Where sea otters are present, they feed on sea urchins and keep their numbers in check. Where otters are missing, sea urchins multiply and eat the kelp until there is almost none left. Without food, the sea urchins then die, and the kelp regrows.

Kelp and Yogurt

What has kelp got to do with yogurt? Seaweeds—raw, cooked, or dried—can make delicious food. They are also rich in vitamins and materials. Not only this, seaweeds contain chemicals called alginates that are used in the food, medical, and cosmetics industries. Alginates form thickeners, binders, and stabilizers for all sorts of items: drug capsules, lipsticks, shampoos, toothpaste, low-fat mayonnaise, and yogurts. Brewers even add them to beer to make the foamy head last longer.

Sea Grass

Sea grasses are the only flowering plants that live underwater in the sea. They are true grasses, with proper stems, roots, leaves, and flowers. They grow best in clear, shallow water that has a soft bottom of nutrient-rich mud or sand. Where they grow well, they form large meadows that support many animals.

Few animals can eat the tough leaves of sea grass, but many use the leaves as a floating platform. If you examine a sea-grass leaf closely, you'll see green slime on the surface—a layer of microscopic algae. Snails, fish, and other animals graze on the algae. Many more creatures—tubeworms and sea squirts among them—attach themselves to sea-grass leaves and filter the passing water for the plankton on which they feed. These small animals, in turn, are eaten by crabs, fish, and other large animals.

In tropical waters, some sea urchins and parrot fish eat sea-grass leaves, as do some endangered animals, such as green turtles, manatees, and dugongs. Manatees and dugong are also called sea cows. They are large mammals that look a bit like friendly walruses, though they are more closely related to elephants. They munch through energy-rich sea-grass roots as well as leaves.

Knotted wrack, a type of kelp, holds itself afloat with tiny air bladders, which act like water wings.

Fact File

▲ The largest giant kelp can grow 12 inches (30 cm) a day.

▲ The total amount of kelp harvested every year in the United States for human use weighs 65,000 tons (60,000 metric tons), or as much as 10,000 African elephants.

▲ There are more than 6,000 types of seaweeds.

Left: Look closer at this delicate frond of seaweed and you'll discover it is in fact one of the world's most bizarre fish, called a weedy sea dragon. Its long, green fins camouflage it among kelp and sea grass.

Under Threat

Sea grasses are under serious attack—from us. In some places, industries pollute seawater with oil or with heavy metals such as lead and zinc. All these pollutants kill sea grasses. Sewage also kills sea grass because it is very rich in nutrients—the nutrients make microscopic algae in the water multiply so much that they turn the water cloudy, depriving the sea grass of light. Fishing boats that dredge the seafloor for shellfish or drag fishing nets through sea-grass meadows also make the water cloudy.

A mysterious disease sometimes destroys sea grass. Scientists disagree on the exact cause of the disease, but some think it might be brought on by an increase in ocean temperatures caused by global warming. If sea-grass plants are already weakened by pollution, rising temperatures might be enough to finish them off.

In some parts of the United States, bays have lost more than half their sea-grass meadows in the last 40 years. Marine biologists are trying to reintroduce sea grasses into areas where they were once abundant, but if global warming is causing the decline, the new plants are unlikely to survive.

Sea-grass meadows are not only important for wildlife. They also bind together mud and sand, and so help protect coastlines from being slowly worn away by the sea.

The dugong is one of the few animals that can eat the tough leaves of sea grasses.

The bright blue flecks on the lips of this giant clam are algae. They produce food from sunlight, and pass some food to the clam. In return, the clam provides protection.

Strange Homes

Coral reefs—those dazzling, multicolored, fish-rich havens of tropical waters—would not be there but for microscopic algae. But where are the algae on the reef? They are inside the animals. Each coral polyp (*see* page 34) contains microscopic algae called zooxanthellae—these can make up more than half the mass of coral polyps on a reef. The polyps give the algae a safe home in the sunlight, and in return, the algae give the polyps a supply of food, produced by photosynthesis. The algae also help the polyps produce their chalky armor. This close relationship, where both partners benefit, is a form of symbiosis (a close partnership between different species). The zooxanthellae are symbiotic algae. Without them, coral reefs would not exist.

The alga–polyp partners build their biggest reefs in clear, shallow waters in tropical seas, where the zooxanthellae can collect the most sunlight. Clear waters are found in downwelling regions where surface waters are poor in nutrients. Here, coral reefs grow particularly well.

Coral polyps are not the only sea creatures with plants inside them. Animals called sponges do the same thing. Giant clams—those that look as though they could grab your leg (but do not)—sport brightly colored bands of algae in their lips, where the two halves of the shell come together.

Life Without Plants

Scientists once thought that all life depended on energy from the sun. Plants and algae absorb the sun's energy to make food, and animals eat them (or one another) to obtain the food. But another way of life is possible.

In 1977, scientists in a submersible on the floor of the Pacific Ocean, more than a mile below the surface, were looking for hydrothermal vents—places on the seafloor where volcanically heated water gushes into the ocean. They found something much more exciting. On the dark seafloor near a vent they came across huge, pale clams and dozens of giant worms—4 feet (1.2 m) long with bright red, feathery gills (right).

Scientists discovered that creatures living close to hydrothermal vents do not need sunlight or plants at all. They get all their food from bacteria that use energy from the chemicals in the vent water to make food. The giant worms and clams have bacteria inside their bodies. They give the bacteria a home and get food in return. The worms and clams grow much faster and larger than their relatives in shallow water.

Pacific Ocean

The Pacific is the world's largest stretch of water. It covers one-third of Earth's surface and contains about half the world's seawater. At its widest it is 11,000 miles (17,700 km) across.

Bora Bora lies near Tahiti in French Polynesia. Like thousands of other islands in the Pacific Ocean, it is an atoll made by coral fringing a collapsed volcanic crater.

1. Japan's Fishing Grounds
People in Japan get half of their protein from fish. Japan's fishing industry is very active.

2. Mariana Trench
This canyon in the ocean floor was once thought lifeless, but its cold darkness is alive with shrimps, worms, sea cucumbers, and microorganisms.

3. Great Barrier Reef
This coral reef system may be the largest structure ever built by living organisms.

4. Emperor Seamounts
Seamounts are underwater mountains. Their tops provide isolated patches of submerged habitat that are cut off from other seamounts by deep ocean. Some seamounts harbor species that live nowhere else.

5. Midway Islands
These sandy islands, lagoons, and coral reefs were the site of a battle in World War II (1939–1945), but are now a U.S. National Wildlife Refuge.

6. Kelp Forests
Seaweed forests formed by giant kelp. The kelp has recovered since a hunting ban allowed the sea otter to return. The otters eat the sea urchins that eat the kelp.

7. Juan de Fuca Ridge
Scientists found life teeming here, around hot vents in the ocean floor.

8. Monterey Bay
A National Marine Sanctuary stetching 400 miles (640 km) along the California coast. People protect this area for the benefit of many species, such as sea lions and gray whales.

9. Gulf of California
This sheltered sea provides feeding grounds for finback whales. Gray whales also come here to give birth.

10. Galápagos Islands
These isolated islands have unique wildlife, such as seaweed-eating iguanas and flightless cormorants.

 Pacific Facts

▲ Mauna Kea is a volcano that rises from the Pacific Ocean floor to form the Big Island of Hawaii. Measured from its base it is 6 miles (9.6 km) tall, which is taller than Mount Everest.

▲ Ferdinand Magellan, the famous 16th-century round-the-world explorer, named the Pacific for its supposedly peaceful nature.

▲ The deepest point on Earth is Challenger Deep in the Mariana Trench, which is 6.9 miles (11 km) deep. A stack of 30 Empire State Buildings built within it would only just reach the sea surface.

11. Galápagos Rift
In 1977, scientists first discovered hydrothermal vents here on the ocean floor. The vents spew out water hot enough to melt lead, but unique life-forms cluster around them.

12. Humboldt Current
The Humboldt, or Peru, Current causes nutrients to well up from the deep and creates one of the richest fishing grounds on Earth. Penguins and seals hunt here, and sharks and tuna come to feast by the million.

Arctic Ocean

East Siberian Sea

Chukchi Sea

Beaufort Sea

ev Sea

RUSSIA

Sea of Okhotsk

Kamchatka Peninsula

Bering Strait

Bering Sea

Aleutian Islands (U.S.)

Aleutian Trench

Kelp Forests

N

Hudson Bay

NORTH AMERICA

6

7

Juan de Fuca Ridge

Emperor Seamounts

Kurile Trench

Sea of Japan

JAPAN

1

4

5

8

Monterey Bay National Marine Sanctuary

Gulf of California

9

Atlantic Ocean

Gulf of Mexico

Izu Trench

Midway Islands (U.S.)

Hawaiian Islands (U.S.)

hilippine Basin

allenger Deep

Mariana Trench

Micronesia

MARSHALL ISLANDS

Caribbean Sea

Galápagos Rift

11

Pacific

2

FEDERATED STATES OF MICRONESIA

NAURU

KIRIBATI

10

SOUTH AMERICA

PAPUA NEW GUINEA

New Guinea

SOLOMON ISLANDS

TUVALU

Polynesia

Cook Islands (New Zealand)

Galápagos Islands

Peru – Chile Trench

(midocean ridge)

STRALIA

Great Barrier Reef

3

Coral Sea

VANUATU

FIJI

SAMOA

TONGA

TONGA Trench

French Polynesia (France)

Tahiti

Bora Bora

Pitcairn Islands (UK)

East Pacific Rise

Easter Island (Chile)

Humboldt (Peru) Current

12

New Caledonia (France)

Southwest Pacific Basin

Ocean

Tasman Sea

NEW ZEALAND

0   2000 miles

0   2000   km

Pacific–Antarctic Ridge (midocean ridge)

ANTARCTICA

Coral Seas

Most of the Pacific is an almost featureless expanse of open ocean that covers the seabed 2.5 miles (4 km) below. But the Pacific has thousands of islands. Most are the tips of volcanoes that have risen from the ocean floor. In warm, clear waters, coral reefs (right) develop around the edges of these islands. Over time, the islands slowly sink, but the coral keeps growing upward toward sunlight. A ring of coral is left at the surface where the island was, forming a coral atoll. The Pacific is famous for them.

Ocean Animals

Animals live at all depths in the oceans, from the surface to the seafloor 6 miles (10 km) down. They range in size from the microscopic to the largest animal that has ever lived on Earth—the blue whale.

The oceans contain a vast amount of living space. They cover nearly three times as much of Earth's surface as does land, but they are more three-dimensional (3-D), with animals able to live at any depth. As a result, there is 6,000 times more living space in the oceans than on land.

A Drifting Life

Animals that live floating near the ocean surface, among the plankton, are called zooplankton. Many are microscopic, but the largest, such as the biggest jellyfish, can be more than 3 feet (1 m) wide, with tentacles 30 feet (9 m) long.

Small and medium-sized zooplankton eat plant plankton (phytoplankton), which they catch in a variety of ways. Animals called salps and larvaceans pump seawater through their bodies, catching the phytoplankton in gel-like nets. Tiny, shrimplike animals called copepods use hairy legs to waft phytoplankton toward their mouth.

Most large zooplankton eat smaller zooplankton. Arrow worms, for instance, are vicious killers, about 4 inches (10 cm) long. They dart after almost any prey smaller than themselves. Jellyfish capture fish and large zooplankton with their stinging tentacles. Not all large zooplankton are predators, though. Among the exceptions are krill— shrimplike animals about 1 inch (2.5 cm) long that feed on phytoplankton. Krill are very common in antarctic waters, where they are the main source of food for many fish, birds, and whales.

Water Babies

Many ocean animals live in the plankton only when young, as larvae, and settle on the seafloor or seashore to grow into adults. Most larvae look completely different from the adults they will become. Barnacle larvae look

 ## How to Float

The trick in living among the plankton is to avoid sinking. Plankton use various techniques to stay afloat. One is to have lots of body projections, such as spines or hairs, to give the body a larger surface area. Another technique is to use droplets of oil or bubbles of gas to buoy up the body. Gel-like animals, such as salps and comb jellies, get rid of heavy chemicals and replace them with lighter ones.

With water to support their bodies, ocean animals have less need for a skeleton than do land animals. Octopuses have no bones whatsoever, allowing them to form their flexible bodies into endless different shapes and squeeze through tiny crevices. They use their suckered tentacles to grasp and smother prey.

Insects of the Sea

The most common animals on Earth are not ants, flies, or any other kind of insect, but tiny shrimplike sea creatures called copepods, which also live in freshwater and wet places on land. Copepods outnumber all other animals on the planet. In their countless trillions, they probably outweigh any other type of animal, too. In the

oceans, they are one of the main sources of food for fish, making them of great ecological importance.

Copepods are members of a group of animals called crustaceans, which also includes shrimps, crabs, and lobsters. There are almost no insects in the oceans, but crustaceans take their place.

A puffer fish, or puffer (below), wards off attackers by inflating itself like a balloon (inset) and making sharp spines stand out on its skin.

Lionfish use venom to defend themselves. If an attacker presses against one of the lionfish's spines, the spine injects a poison so strong that it can kill a person.

like fleas, some crab larvae look like miniature UFOs, and certain sea urchin larvae look like snowflakes. For animals that spend their adult lives on the seafloor or shore, having planktonic larvae is a good way of spreading to new areas because the larvae can drift great distances on sea currents.

Near the coast, at certain times of the year, planktonic larvae outnumber the normal residents of the zooplankton. Fish and other larger creatures gather together to take advantage of this bumper harvest.

Fishy Business

The true kings of the ocean are fish. More than half of all living species of vertebrates (animals with backbones) are fish, and more than two-thirds of fish species live in the sea—at least 15,000 species. Fish are amazingly successful. They come in all shapes and sizes, from gobies less than half an inch (about 1 cm) long to whale sharks reaching 60 feet (18 m)—as big as whales. Fish live at all depths to about 20,000 feet (6,000 m) down. Some can even breathe air and live on land or glide short distances over the waves.

Scientists classify fish into three main groups: those without proper jaws (lampreys and hagfishes); those with jaws and skeletons of cartilage (sharks and rays); and those with bony skeletons. The majority of fish belong to the bony group, including herring, cod, salmon, tuna, and flatfish such as sole.

Though a few fish graze algae, most fish are predators—agile swimmers with sharp senses and strong jaws for snapping up food. Fish prey on everything from plankton to large sea animals, and many are skilled hunters. Some stun their prey with electricity, others hide and ambush victims, and many use speed and

agility to snatch prey from the water. Some of the sea's top predators are mammal- and fish-eating sharks—the most lethal fish on Earth.

Shark Attack

Sharks have a bad reputation, but not all of them deserve it. Of the 400 or so species of sharks, about 40 species are known to have attacked people. In 2000, there were 79 unprovoked shark attacks recorded worldwide; ten of the people involved died. In comparison, fishers caught millions of sharks in the same year and killed nearly all of them. Sometimes, fishers hunt sharks on purpose, but usually the sharks are unwanted, caught among tuna or other commercial fish.

Sharks use a range of senses to track down their food. They can smell blood from more than 1 mile (1.6 km) away but use sight to identify victims nearby. Like most fish, they can feel the vibrations that travel through water, giving them a sense of touch at a distance. And from close range, they can detect the field of electricity generated by their victim's muscles.

Three species of sharks are responsible for most recorded attacks: the great white shark, the tiger shark, and the bull shark. There are several reasons why these sharks might attack people. In some cases it is a case of mistaken identity—great whites occasionally attack surfboards, probably because they mistake the outline of surfer and board for that of a seal. In other cases a shark might attack if it feels threatened. Only rarely do sharks deliberately attack people in order to eat them.

In fact, you are much more likely to be run over by an automobile on the way to a beach than be killed by a shark when you get there. Nevertheless, swimming in waters where people have reported dangerous sharks is not a good idea.

Coral Extravaganza

In the warm waters of the tropics are the most complex wildlife communities in the ocean: coral reefs. A single coral reef in the Pacific can contain more than 100 species of corals and several thousand species of

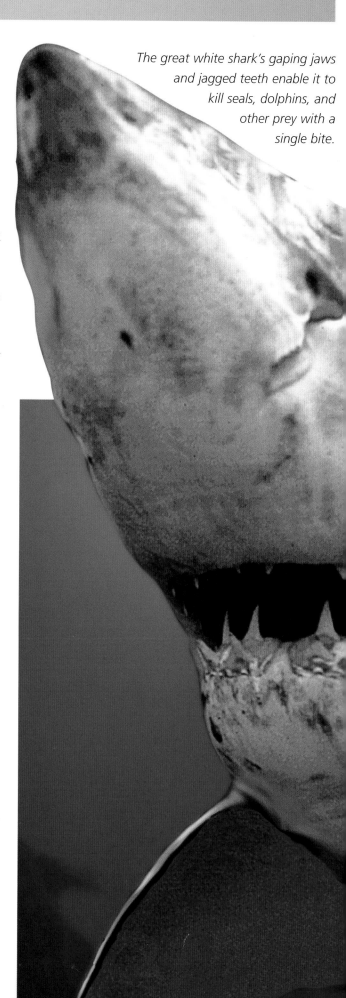

The great white shark's gaping jaws and jagged teeth enable it to kill seals, dolphins, and other prey with a single bite.

Many corals live in large groups, sharing a skeleton and linked by sheets of tissue. The individual star-shaped animals, called polyps, emerge from the skeleton at night to catch plankton.

invertebrates (animals without a backbone), while providing a hunting ground for hundreds of fish species and other animals.

The source of this abundance is the humble coral polyp, a tiny, tentacled animal that looks a bit like a flower. Coral polyps use their mobile tentacles to strain seawater for plankton, but they also obtain food from the algae that live inside their bodies (*see* page 27). Some polyps live on their own, but most clone themselves to form huge colonies in which all the individuals are joined together. Each coral species builds a colony of a particular shape— they might look like brains, mushrooms, antlers, or giant plates. Corals

Arrowhead crabs live among the nooks and crannies of coral reefs. They use pincerlike claws to snip chunks of flesh off coral polyps and other animals.

sport all the colors of the rainbow—and the reef's other colorful inhabitants add to this palette, making a living reef an explosion of fascinating shapes and colors.

Many corals secrete a carbonate skeleton to protect themselves and support the colony. When the corals die, the skeletons remain in

Bright colors are often a sign of danger. The frilly skin of many sea slugs contains microscopic stings, stolen from the sea anemones on which the slugs prey.

35

Coral Facts

▲ A barrier reef is a structure built by coral that lies parallel to the shore, on the far side of a lagoon.

▲ The largest coral reef system is the Great Barrier Reef of Australia. It is 1,250 miles (2,000 km) long and clearly visible from space. It is made up of more than 2,500 separate reefs.

▲ The biggest coral atoll is Kwajalein in the Marshall Islands of the western Pacific. It is more than 44 miles (70 km) long and 20 miles (32 km) wide.

▲ The soft white sand of many tropical beaches forms from coral skeletons broken down by the sea.

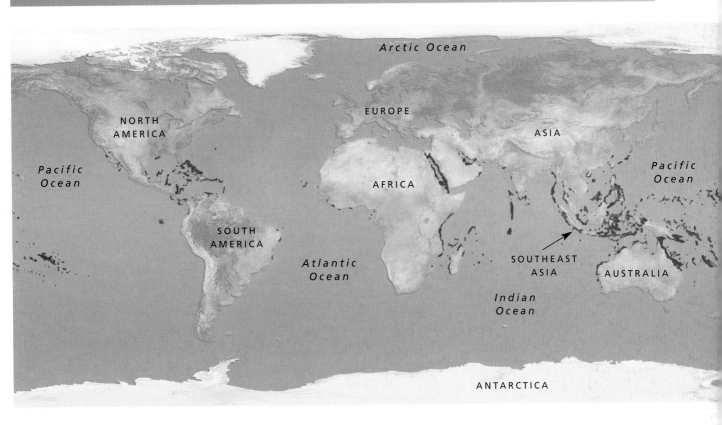

place and gradually build up to form the reef. A coral reef's nooks and crannies provide havens where seaweeds and animals can settle and hide. Most reef animals graze the thin covering of algae that grows on the reef, or, like the coral polyps, they strain the water for plankton. Only a few animals, such as parrot fish and some starfish, can eat corals.

Some reef animals have their food delivered on a living plate. On many reefs there are cleaning stations where fish line up to have their skin, mouth, and gills cleaned by fish and prawns that they might otherwise eat. The visitors seem to benefit by having parasites removed and wounds cleaned.

World Without Light

More than about 3,300 feet (1,000 m) down, the sea is totally dark, except for light made by the animals themselves, or the headlights of submersibles carrying marine biologists.

For animals living in the dark zone, senses other than sight come to the fore. Most scavenging deep-sea fish find food by smell.

Coral reefs (red) flourish in sunny, clear, shallow water throughout the tropics, where the sea is warm. They grow near the coast and on underwater mountaintops.

Chemical Weapons

Many coral reef animals live fixed to one spot. They can't run away when danger threatens, so they need other ways of defending themselves. A popular strategy is to make chemicals that taste repulsive or are poisonous to predators.

Some of these chemicals have proved to be useful as drugs for treating cancer and other diseases. One example is ecteinascidin-743, which is currently being tested as an anticancer drug. It comes from a type of sea squirt, an invertebrate, living on coral reefs in the Caribbean. Another possible drug is discodermalide, which is being tested to treat leukemia (blood cancer). Discodermalide comes from a type of sponge, also from the Caribbean.

37

and viper fish can partially dislocate their jaws to swallow large meals. Many of these fish have a stomach that stretches, allowing them to swallow prey bigger than themselves.

Those hunting live prey can also feel the telltale ripples and vibrations in the water that other animals make. The long, sensitive tendrils of some deep-sea anglerfish and the elongated fins on tripod fish's heads are suited to this purpose. Some fish can even detect the tiny electrical signals produced by their victims' muscles.

Anglerfish have another trick up their sleeve. Instead of hunting down their prey, they use a glowing lure, attached to a long stalk on the head, to draw victims toward them. The lure looks like a wiggling worm—an irresistible temptation to hungry animals. When a victim comes into striking range, the anglerfish snatches it in its massive jaws.

Many deep-sea creatures depend for food on the snowfall of dead animals, body parts, and feces from high above. Even so, food is very scarce, and animals must conserve their energy. Deep-sea fish move very little, and their muscles are small and puny compared to those of other fish. But they make up for this by the size of their mouth and stomach. Gulper eels, or deep-sea swallowers, have vast jaws and a snakelike body,

Finding a Mate

Besides being pitch-black, the deep-sea world is enormous and mostly empty. Because food is hard to find, animals are scattered much farther apart than in surface waters. The darkness and emptiness can also make it difficult for fish to find mates. Many probably use their sense of smell to zero in on scent chemicals (pheromones) produced by a potential mate. At close range, anglerfish can identify suitable mates by the pattern of light from the glowing lure, and other fish species probably recognize light patterns, too.

Once an animal has found a mate, it is all too easy to lose it again in the inky blackness. Some male anglerfish have a drastic solution to this problem: They sink their teeth into the female's body and stay attached for the rest of their lives. In such species, the male is much smaller than the female. The female's skin grows over the male, and he becomes a parasite, feeding on her blood.

Animals of the deep have to catch whatever they find. Viper fish (far right) have monstrous, cagelike teeth to trap hatchet fish (above right) and other prey.

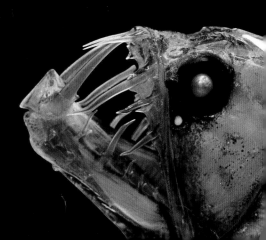

Making Light Work

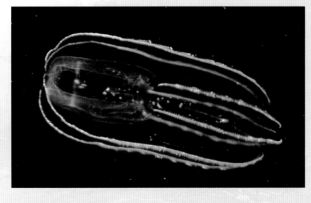

Many animals in the twilight zone, and some in the dark zone, can produce their own light (bioluminescence). This comb jelly, for instance, produces shimmering, rainbow-colored lights along its transparent body. Bioluminescent light comes from a chemical reaction in special light organs and is usually blue or green. It has a variety of uses. Some deep-sea squids squirt out a cloud of shimmering light that distracts an attacker, giving the squid a few seconds to escape. Anglerfish use a luminous lure to catch prey, and lantern fish sport complex patterns of lights that help them recognize other members of their species. Some fish of the twilight zone have lights on their underside. Seen from below, the spots of light help camouflage them against the faint glow of the sea surface.

Mysteries of the Deep

Even today, there are probably many undiscovered animals in the oceans. With the ocean so difficult to study, and with so much water to search, it is not surprising that animals, even large ones, remain to be seen.

Marine scientists only encountered the weird animal communities of deep-sea vents in 1977. In 1997, they discovered something even weirder: worms living inside frozen methane gas on the floor of the Gulf of Mexico. And one species of whale—the True's beaked whale—was only found alive at sea in July 2001. A male leaped 24 times out of the water near a ferry in the Bay of Biscay, France. Luckily, there were whale experts on board to witness and photograph the event.

For centuries, most scientists thought the giant squid was a myth. But in the 1850s, a Danish zoologist decided to examine the reports of sightings and analyze samples. He soon confirmed that the creature was real—it was a giant version of the common squid.

Now that more scientists are studying the ocean, we can expect more surprises. Despite teams of researchers searching for it, no one has seen a giant squid alive in its midwater habitat. In the 1990s, however, scientists descended to 1,000 feet (300 m) to study another animal, called the vampire squid. They had thought it was a weak-muscled,

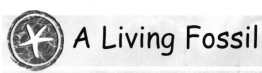

A Living Fossil

In 1938, fishers caught a strange, stocky fish, about 5 feet (1.5 m) long, off eastern South Africa. The skipper took the fish to the local museum curator, Marjorie Courtenay-Latimer, who recognized it immediately. Nobody believed her when she said it might be a coelacanth—a fish believed to have been extinct for 60 million years. Luckily, she had the foresight to have the fish stuffed, and when fish expert J. L. B. Smith came to examine it, he could hardly believe his eyes. The coelacanth's internal organs had been thrown away when it was stuffed, so Professor Smith began searching for an intact specimen—a hunt that took another 14 years. In 1998, coelacanth enthusiasts discovered a second species—this time off Indonesia. Who knows what other large sea creatures remain undiscovered?

Ocean Discoveries

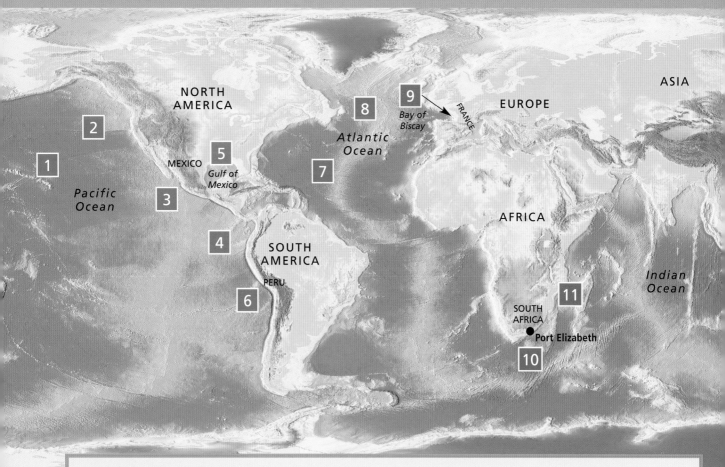

The oceans are so vast, deep, and unexplored that all sorts of mysterious creatures might lurk in their depths, awaiting discovery. The finding of animals such as giant squid, megamouth sharks, and unknown whales continues to amaze experts. Some of these creatures were previously known, but only from long-dead fossils.

1. Megamouth Shark
In 1976, this large-mouthed shark was found entangled in a parachute in the sea near Hawaii. It feeds on plankton near the sea surface at night, but descends to the depths during the day, which explains why it is so rarely seen.

2. Six-Eyed Spookfish
This fish looks like a pike but has six eyes. It was discovered in 1958, at a depth of more than 330 feet (100 m).

3. Monoplacophoran
Scientists dredged this shelled mollusk, which looks like a giant limpet, off Mexico's west coast in 1952. Biologists thought its kind had died out 350 million years ago.

4. Vestimentiferan Worm
In 1977, scientists discovered hundreds of huge tube worms, up to 7 feet (2 m) long, among dozens of other new species of fish, clams, shrimps, and crabs at deep-sea vents.

5. Ice Worm
In 1997, scientists discovered this new species of worm living in methane ice on the floor of the Gulf of Mexico.

6. Lesser Beaked Whale
Parts of this whale's body were first found near a fish market in San Andres, Peru, in 1976. Several complete specimens were later found, and it was officially described as a new species in 1991.

7. Vampire Squid
This tentacled, purple-black animal was caught in an Atlantic deep-sea sample in 1903. In 1946, scientists confirmed it is neither a true squid nor an octopus, but a member of a related group thought to have been extinct.

8. Giant Squid
The species that lives in the North Atlantic grows to 60 feet (18 m) long. It has never been seen alive.

9. True's Beaked Whale
The first confirmed sighting of a living True's beaked whale took place in July 2001 in the Bay of Biscay, off France.

10. Deepwater Stingray
The first specimen washed up on a beach at Port Elizabeth, South Africa, in 1980.

11. Coelacanth
The first live specimen known to science was caught near South Africa in 1938. At that time, the only other coelacanth specimens were fossils as old as dinosaurs.

sluggish animal, so they were amazed to see it swimming at high speed, rapidly changing shape, and turning in tight circles. The scientists had to rewrite their textbooks.

Marine Migrations

Many marine animals make incredibly long journeys to find food or to breed. These epic voyages are called migrations.

Most migrations are seasonal. At a certain time of year, the animal migrates to a new home, where conditions are better for the next stage in its life cycle. It might migrate to breed, for instance, but then return later in the year to somewhere better for feeding.

Salmon and freshwater eels not only migrate through the sea but also travel up rivers. This is a challenge, because freshwater contains far smaller concentrations of chemicals than salt water. The move requires changes in the animal's body function.

Several species of Pacific salmon lay their eggs in the gravel-covered beds of North American rivers. When the baby fish hatch, they set off downriver, and grow into adults by the time they reach the ocean. Years later,

Living in Sand

At the very edge of the ocean is the shore, where conditions for animals such as this clam change drastically: the seabed is exposed twice a day by the tide. Unlike rocky shores, where hardy sea life survives attached to rocks, sandy shores often appear devoid of life. But sandy beaches can be full of animals—beneath the surface. Most sandy-beach animals burrow, especially at low tide, to avoid predators and to avoid drying out. Many clams stay buried at high tide, too, but extend their siphons (tubes) into the water. The siphons pump the water and filter out food particles.

siphons

the adults return to breed in the exact place where they hatched. To get there, they must swim upstream through rapids and leap up waterfalls. They must avoid all sorts of natural and artificial obstacles—dams, fishing nets, anglers, grizzly bears, and pollution.

When they arrive at the breeding grounds, they scoop out a gravel nest. The female lays her eggs, a male fertilizes them, and then both male and female die.

Atlantic eels make a reverse journey. Adults lay their eggs on the floor of the Sargasso Sea—a still patch of the Atlantic Ocean near Bermuda. The young eels rise to the surface, drift on ocean currents, and swim into North American and European rivers. They mature in lakes and rivers, and then return to the Sargasso Sea to breed.

The endangered hawksbill turtle roams the tropical oceans. When it is time to mate and lay eggs, it travels to one of only a handful of nesting sites worldwide.

Many seabirds, sea mammals, and turtles make journeys just as long as those of eels and salmon—or even longer. The Arctic tern flies an astonishing 20,000 miles (32,000 km) each year on its round trip between the Arctic Ocean and Antarctica. It breeds on the Arctic coast in summer, but flies south in autumn to spend the rest of the year in the ocean around Antarctica,

Although sea lions need to rest, mate, and give birth on beaches, they move much more easily in the ocean.

Spinner dolphins can swim at 25 mph (40km/h). They hunt in packs, rounding up small fish such as herring into tight shoals before moving in for the kill.

where it is summer at this time. As a result, it probably sees more hours of daylight in a year than any other animal on Earth.

The gray whale breeds in the Gulf of California in winter, away from the attentions of killer whales (orcas), which hunt in cooler waters. In summer, adults and calves swim north to the Arctic, where food is more abundant. In total, the whales swim up to 12,500 miles (20,000 km) each year— the longest migration of any mammal.

Somehow, green turtles born on Ascension Island find their way back as adults to this tiny island in the Atlantic Ocean. They travel on a 1,370-mile (2,200-km) journey from feeding grounds close to Brazil to mate in the waters off Ascension. The females then struggle ashore to lay their eggs.

Marine Mammals

Fish may be the kings of the oceans, but the biggest sea creatures are whales. Whales belong to a class of animals called mammals, which also includes dogs, bats, elephants, and humans. Mammals are warm-blooded animals that breathe air and feed their young on milk. They evolved on land about 210 million years ago, but about 65 million years ago, some began living in the sea.

The first marine mammals were fierce meat eaters that looked a bit like crocodiles. Over millions of years, these crocodile-like mammals became streamlined, their stumpy legs became fins, and their hair gradually disappeared. Although they still have to come to the surface to breathe, they have now mastered the oceans. They are the cetaceans: whales, dolphins, and porpoises.

Some cetaceans—the toothed whales—hunt fish and squid. These cetaceans are fast and agile swimmers, and have teeth for gripping their prey. Dolphins, killer whales, porpoises, and sperm whales are all toothed whales. Some toothed whales have large brains and may be highly intelligent. Bottle-nosed dolphins may even solve problems by thought, and they sometimes use tools. In the shallow seas off the coast of western Australia, bottle-nosed dolphins hold a sponge in their mouth to protect their snout while they dig for shellfish.

Other cetaceans—the baleen whales—have a giant sieve (a baleen plate) inside the mouth instead of teeth. These whales take massive gulps of seawater and squeeze it through the sieve to collect small animals, such as krill. Baleen whales are the giants of the ocean. The blue whale, which weighs up to 209 tons (190 metric tons) and reaches up to 110 feet (33 m) long, is probably the largest animal that has ever lived on Earth.

Some baleen whales sing to one another to communicate. Humpback whales are famous for their long, haunting songs and their cooperative hunting techniques. In the breeding season, adult males sing a 20-minute song to attract females and to intimidate other males. In polar waters, groups of humpbacks blow curtains of bubbles to herd schools of fish and krill into a tight ball. Then they lunge upward through the close-packed school—a bumper harvest for a minimum amount of effort.

Humpback whales, although among the largest of animals, eat only tiny prey—krill and small fish. In one mouthful, they swallow many thousands.

Indian Ocean

The Indian Ocean formed about 125 million years ago when Africa tore free from Antarctica and drifted north. Now this ocean's wonders include coral islands, Antarctic breeding grounds, and giant river deltas.

5

Snorkelers are drawn to the clear turquoise waters and abundant coral reefs of the Maldives. There are vacation resorts on more than 70 of the 1,300 Maldive islands.

Monsoons

In the northern Indian Ocean, the winds nearly always blow from the ocean to the land in summer, bringing very wet weather to the land. In winter they reverse, blowing from the land and making the weather hot and dry. Winds blowing in this seasonal pattern are called monsoons, and they have a profound influence on the climate of the countries around the Indian Ocean.

The wind direction is so constant that dhows, the ancient sailing vessels of the northern Indian Ocean, do not come equipped with adjustable sails. The sailors must lower and reattach the sail when the wind changes direction or the boat needs to sail on a new course. In other parts of the world, sails are readily adjustable because sailors are used to changeable winds. The dhow to the right is sailing to the island of Zanzibar off east Africa.

Fact File

▲ The Portuguese explorer Vasco da Gama sailed across the Indian Ocean in 1497, after rounding the southern tip of Africa. Ever since, the Indian Ocean has been a trade route for silks, rugs, tea, and spices.

▲ The yellow-bellied sea snake swims throughout the Indian Ocean and into the Pacific. But like other tropical life in the Indian Ocean, it can't reach the Atlantic due to the cold sea around southern Africa.

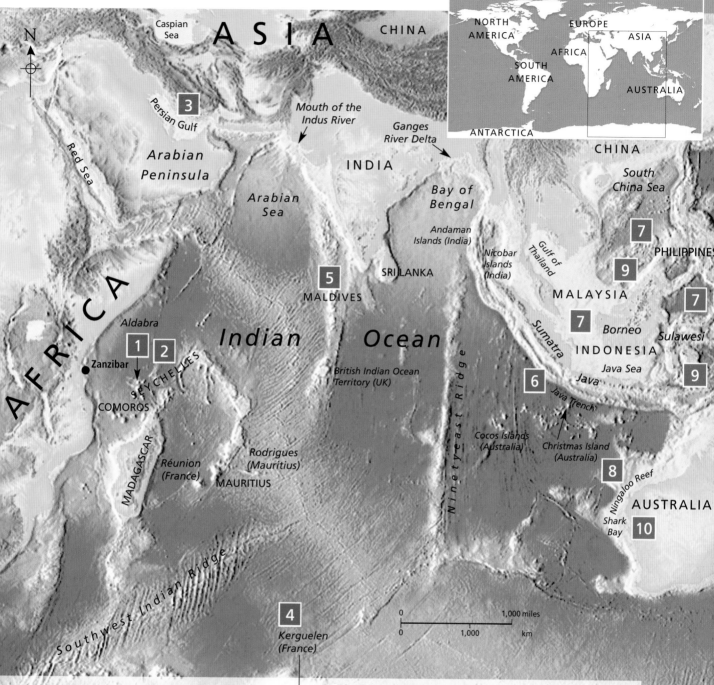

N

Caspian Sea

ASIA

CHINA

Persian Gulf

3

Mouth of the Indus River

Ganges River Delta

Arabian Peninsula

INDIA

Red Sea

Arabian Sea

Bay of Bengal

Andaman Islands (India)

Nicobar Islands (India)

Gulf of Thailand

South China Sea

CHINA

7

9

PHILIPPINES

AFRICA

Aldabra

Zanzibar

1 2

SEYCHELLES

COMOROS

MADAGASCAR

Réunion (France)

Rodrigues (Mauritius)

MAURITIUS

Indian

5

MALDIVES

SRI LANKA

Ocean

British Indian Ocean Territory (UK)

Ninetyeast Ridge

MALAYSIA

7

Sumatra

Borneo

INDONESIA

Java Sea

Java

Java Trench

6

Sulawesi

7

9

Cocos Islands (Australia)

Christmas Island (Australia)

8

Ningaloo Reef

AUSTRALIA

Shark Bay

10

Southwest Indian Ridge

4

Kerguelen (France)

0   1,000 miles
0   1,000   km

NORTH AMERICA

EUROPE

ASIA

AFRICA

SOUTH AMERICA

AUSTRALIA

ANTARCTICA

1. Aldabra

A coral atoll that is a U.N. World Heritage Site. It is home to the Aldabra giant tortoise as well as huge robber crabs and land hermit crabs.

2. Seychelles

These palm-fringed islands are fragments left behind when India broke free of Africa less than 200 million years ago.

3. Persian Gulf

Earth's largest oil reserves lie on the shores of this sea and under the seabed.

4. Kerguelen

Many mammals and birds of the southern Indian Ocean breed here. Albatrosses, seals, and penguins abound.

5. Maldives

These low-lying islands are the tops of submerged volcanoes. Growing coral reefs keep them above water, but nowhere are they higher than 6 feet (1.8 m).

6. Java Trench

This submarine canyon is 4.6 miles (7.45 km) deep, the same as 19 Empire State Buildings.

7. Cyanide and Dynamite Fishing

Population growth in Southeast Asia has led to destructive fishing methods around the coral reefs of Indonesia. Fishers kill fish for the food market with dynamite, and use cyanide to stun fish for the pet trade.

8. Ningaloo Reef

Warmed by a tropical ocean current, this long coral reef fringes the coast of Australia. Visitors come to watch migrating whale sharks, the world's largest fish.

9. Indo-West Pacific

This system of small seas among the islands of Indonesia contains a rich diversity of sea life. Tens of thousands of species of animals, plants, and microorganisms from both the Indian and Pacific oceans mingle in this area.

10. Shark Bay

The world's largest sea-grass bed, more than 75 miles (120 km) long and home to thousands of dolphins and sharks, as well as 10,000 dugongs, a kind of sea cow.

Oceans and People

Many thousand of years ago, our ancestors used the oceans only as a source of food. Later they became proficient sailors and crossed the seas to explore and colonize the whole world.

The first long-distance sailors may have been the people who colonized the islands of the western Pacific Ocean. Around 6,000–7,000 years ago, people sailed east from New Guinea to the Solomon Islands, and by 3,500 years ago they had reached the middle of the Pacific Ocean. To get there, they would have had to sail hundreds of miles beyond the sight of land.

While Pacific islanders were bravely setting out across the ocean, the seafarers of Africa and Europe were much more cautious, and generally sailed close to the coast. Even so, they made some daring voyages.

According to the Greek historian Herodotus (484–425 B.C.), explorers from Phoenicia (now Lebanon) sailed all the way around Africa in about 600 B.C. Helped by regular stops to obtain provisions, the sailors took three years to make the epic voyage.

Among the greatest early seafarers of Europe were the Vikings—the people who lived in Scandinavia between the years 500 and 1100. In the 8th century A.D., Vikings attacked towns in Britain, Ireland, and France, spreading fear throughout the countryside. But some Vikings came to farm and trade and not to pillage.

In about the year 1000, Vikings set off west from Greenland. Eventually they reached a forested coast that they called Markland, which means "land of woods." Today, historians believe this to be southern Labrador, Canada—which means that Vikings discovered the Americas nearly 500 years before Christopher Columbus did.

Age of Exploration

Starting in the 15th century, a series of Portuguese sailors set off on major voyages to set up new trade routes and find new lands to colonize. Historians call this period the Age of Exploration. In 1486, Portuguese sailor Bartholomeu Dias became the first known European to sail around the tip of Africa from west to east. In 1498, Vasco da Gama sailed from Portugal to India, and in doing so, opened up sea trade between Europe and Asia.

One of the world's most influential sea journeys took place in 1492. Italian explorer Christopher Columbus led a

Left: The Phoenicians built a sea empire by controlling trade routes across the Mediterranean Sea. Their ships and sailors were the best in the Mediterranean.

Spanish-sponsored expedition to find the western route to the East Indies (Southeast Asia). After five weeks, his three ships landed on an island in the Bahamas—Columbus had crossed the Atlantic. On his heels came many other expeditions, with John Cabot from England reaching mainland North America in 1497.

It might seem that ocean travel is less important today than it used to be, but nothing could be further from the truth.

More than 90 percent of bulk goods—everything from fruit to refrigerators—travels in the hold of cargo ships rather than in aircraft. In your household, many of the foods and domestic items (or their parts) traveled by sea at some stage.

Undersea World

Explorers have now mapped every square mile of Earth's land surface, but they have barely begun to explore the mysteries of the undersea world.

One of the oldest ways of making use of the sea is spearfishing. It is still common in parts of the tropics, including this bay in Madagascar. Besides spearing fish, the fishers cast small nets or look for squid and crabs while wading through the water.

Newfoundland
Bristol
EUROPE
NORTH AMERICA
Nova Scotia
Lisbon
Salas
Bahamas
Cuba
Hispaniola
AFRICA
SOUTH AMERICA

Around the World

After pioneering voyages of the 15th century by explorers such as Columbus, the time was ripe for someone to try to sail all the way around the globe. In 1519, Portuguese nobleman Ferdinand Magellan left Spain in five small ships (right). His aim was to sail west to the Spice Islands (the Moluccas in Indonesia) to claim them for Spain.

Magellan assumed that the islands were not far beyond the Americas— he had no idea that the world's biggest ocean, the Pacific, lay in the way. During the 100-day crossing of the Pacific, the crews were reduced to eating rats and mice, chewing leather, and drinking soup made with wood shavings. When the sailors arrived in the west Pacific, they met an unfriendly reception. Islanders stole their provisions, and Magellan himself was killed by islanders in the Philippines.

Two ships did reach the Spice Islands in November 1521. But only one made it back to Spain, arriving in September 1522. Of the 237 men that had set out on the expedition, just 18 completed the amazing around-the-world trip.

→ Bartholomeu Dias, 1486
→ Christopher Columbus, 1492
→ John Cabot, 1497
→ Vasco da Gama, 1498
→ Ferdinand Magellan, 1519–1521

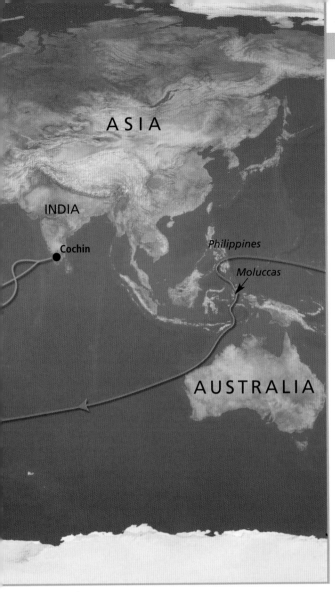

ASIA

INDIA

Cochin

Philippines

Moluccas

AUSTRALIA

scientist Aristotle described a type of diving bell that used metal breathing pipes. It is not clear exactly how, or if, the device worked.

The first good description of a working diving device dates from around 1700. At this time, English scientist Edmond Halley designed a diving bell made of wood. It enabled divers to recover objects from shipwrecks. Helpers sent fresh air down in barrels, while divers explored underwater and returned to the bell at short intervals.

In the 19th century, diving suits became a better option than diving bells. In 1837, the German Augustus Siebe invented the metal helmet and diving suit that you see in old Hollywood adventure movies. Divers using this equipment could descend much farther (200 feet, or 60 m) and with much greater mobility than those using diving bells.

Scuba Diving
In 1943, two Frenchmen—explorer Jacques Cousteau and engineer Émile Gagnan—revolutionized diving with a new invention:

Replicas of Columbus's ships sail past Florida's Kennedy Space Center in 1992 to mark the 500th anniversary of Columbus's first voyage across the Atlantic.

The earliest records of diving come from Mesopotamia (now Iraq). About 6,500 years ago, Mesopotamian divers were holding their breath for two minutes and plunging 33 feet (10 m) below the surface to gather shellfish. In the 4th century B.C., the Greek philosopher and scuba (self-contained underwater breathing apparatus). Thanks to scuba, divers are now free to explore underwater without being encumbered by breathing tubes connected to the surface. However, because the body can't safely absorb nitrogen or oxygen gas at very high pressure, scuba diving is limited to depths of about 200 feet (60 m) with air, or 610 feet (185 m) when breathing a mixture of oxygen and helium. To dive beyond this, some divers have tried filling their lungs with oxygen-rich liquid to prevent the pressure crushing the lungs. During their dive, they continue to breathe the liquid, their lungs acting like gills.

Submarines and Submersibles

A Dutch inventor named Cornelis Drebbel is thought to have built the world's first working submarine. Drebbel's craft was a completely enclosed wooden boat, covered with greased leather to make it waterproof. Oars extended through tight-fitting leather flaps in the sides and provided the only means of propulsion. Between 1620 and 1624, Drebbel took his submarine on test runs in the Thames River in England, rowing it along about 13 feet (4 m) below the surface. He is said to have taken King James I for a ride in it. People now use a variety of underwater

Right: Before the invention of scuba gear, divers had to wear very bulky suits and stay attached by hoses to air tanks on the surface.

Below: Today, both snorkelers and divers use lightweight masks and wetsuits and plunge into the oceans for fun—but diving remains a dangerous sport.

50

craft to explore the oceans. Scientists increasingly use *remotely operated vehicles* (ROVs), which don't contain a crew and can be piloted from a ship. ROVs remain attached to a mothership by cables, but the latest submersibles—called *autonomous underwater vehicles* (AUVs)—are completely free moving. These small, robotic craft can be programmed to stay underwater for long periods, carrying out research.

Ocean Resources

The oceans are full of valuable resources. Many, such as the minerals on the bottom of the sea, are difficult and expensive to gather. But more accessible resources, such as fish, have been massively overexploited in most parts of the world and some fish stocks are in serious decline.

Water itself is a valuable commodity. In the United States, the average person consumes about 600 quarts (570 liters) a day for drinking, cleaning, and flushing the toilet—about 200 times the amount used by a person in Equatorial Guinea in Africa.

Seawater is too salty for drinking or watering crops. However, in places where there is little freshwater, desalinization facilities can be used. These use sunlight to turn seawater to water vapor. When seawater evaporates, it leaves its salt behind. The desalinization plant collects the freshwater vapor and turns it back to liquid.

Worldwide, the oceans provide less than a tenth of people's food by weight. But this fraction is important, because seafood is a major source of protein in tropical countries, where other protein sources are scarce. Oily fish, such as salmon and mackerel, are very healthy to eat. These fish are rich in certain fats that help lower people's blood cholesterol levels. Lower cholesterol means a reduced risk of heart disease and stroke.

Into the Abyss

The deepest point in the oceans is the Mariana Trench—a vast canyon on the floor of the Pacific Ocean. In 1960, Frenchman Jacques Piccard and American naval lieutenant Don Walsh descended to the bottom of the Mariana Trench in a submersible called *Trieste*. Their craft consisted of a ball-shaped metal compartment slung below a submarine-shaped tank filled with gasoline, for buoyancy. After 5 hours, the craft touched bottom at 6.78 miles (10.9 km). The trench is at least 6.86 miles (11 km) deep, and uncrewed craft might dive even deeper in the near future.

These days, submersibles such as *Deep Rover* (right) are kept busy by scientists eager to explore the unknown deep sea, and by engineers inspecting cables and pipelines on the seabed.

Fished to the Limit

Each year, catches of certain marine fish get smaller, although fishers keep trying harder to catch them. Fishers remove more fish from the oceans than are replaced by breeding. As a result, many of the world's fish supplies are in decline.

These trawlermen in New Brunswick, Canada, have just hauled aboard a catch of herring. Unlike popular fish such as cod, local herring are not yet overfished.

stocks to exploit. So the cycle continues. The United Nations recently estimated that 13 of the world's 17 most important fishing areas were either overfished or fished to the limit.

One answer to the overfishing problem is fish farming. Salmon and trout farms have already lowered the price of these fish drastically, and the fish are no longer luxury items. Some fish farmers rear fish in net cages suspended in the sea, while others build seawater ponds close to the shore. Today, fish farming accounts for about a quarter of all marine produce. However, there are drawbacks.

Keeping fish crowded together encourages fish diseases to prosper, so farmers have to dose the fish with antibiotics. Parasites, such as fish lice, also thrive in the crowded conditions. Some scientists are concerned that fish lice from farmed salmon are devastating natural populations of salmon in the northeast Atlantic.

The wastes produced by farmed fish also affect the environment. They contain nutrients that make phytoplankton grow rapidly, altering the balance of the natural biological community.

Governments know that overfishing is causing serious problems, but they cannot agree on how to manage fishing. The industry is a bit like a gold rush, with different countries catching as many fish as they can before the supplies run out. In the process, fish populations are decimated. When there are no more fish left to catch, the fishers lose their jobs or hunt elsewhere for new fish

Minerals from the Sea

The salt in seawater is another valuable commodity. Ever since people started farming and eating a lot of cereals and vegetables, they have used salt in their diet. People still gather salt by channeling seawater into shallow ponds where the hot sun turns the liquid to vapor, leaving the salt behind.

Left: In Vietnam, people let seawater evaporate and collect by hand the salt that is left behind. This woman first piles up the salt, then collects it when it is dry.

Drilling rigs can reach oil deep beneath the seabed. Natural gas (methane) often comes up with the oil, so the gas is burned off to prevent an explosion.

Even more important than solid minerals are the oil and gas supplies that come from reserves under the seafloor. Oil and gas drilled from below the sea provide more than 40 percent of the world's supply of these fuels. As supplies from land get used up, oil and gas exploration moves into ever deeper waters.

Most of the other chemicals in seawater are in amounts too small to be worth harvesting, even though some of these substances are valuable. In some places, minerals such as gold and diamonds are eroded from rocky shores by the sea. They are sifted and sorted naturally by waves and currents to settle near the shore, where they are mined. Surprisingly, however, people make more money from sand and gravel than from any other solids mined from the sea. House and road builders use these materials for making concrete and for many other purposes.

Today, tourism and leisure are among the most important ways of using the oceans. Surfing, diving, sailing, and fishing have all become multimillion dollar industries.

⚡ Resource Facts

▲ Between the 1600s and 1980s, whaling reduced the populations of the largest whales, including blue whales, to a tenth of their former number.

▲ There is enough gold dissolved in the oceans to provide everyone in the world with a lump weighing about 9 pounds (4 kg).

Arctic Ocean

At the top of the world is the Arctic Ocean, surrounded by continents on all sides. Ice formed by the freezing of seawater covers almost the entire Arctic Ocean in winter, and only powerful ice-breaking ships can pass through.

 ## Arctic Crossing

As late as the 1890s, many geographers thought that a continent lay at the North Pole, or at the very least, that the Arctic was a shallow sea. Norwegian scientist and explorer Fridtjof Nansen was one of the few who did not. Nansen designed the *Fram*, a boat that would rise up out of the ice rather than be crushed by it. Between September 1893 and August 1896, the *Fram* drifted with the Arctic ice, at one point coming to within 250 miles (400 km) of the North Pole. Nansen and a companion set out for the Pole on foot but had to turn back in severe weather and spend winter on the ice. Nansen's courageous expedition proved that there was no major continent in the Arctic, just ocean. His team made soundings through the ice and showed that the ocean was more than 6,500 feet (2,000 m) deep in places.

When the great glaciers that flow off the Greenland ice cap meet the ocean, large chunks of ice, or icebergs (left), break off and float away. Once free, the icebergs drift with ocean currents and the wind, and can end up far south in the Atlantic. Some even reached Bermuda in 1907 and 1926.

 ## Fact File

▲ Even in summer, half the Arctic Ocean is covered in sea ice, much of it more than 10 feet (3 m) thick.

▲ Every year at the North Pole, six months of darkness is followed by six months of light.

▲ In places where the Arctic Ocean is free of ice cover, winter storms cause thorough mixing of shallow waters. The mixing stirs up nutrients, which leads to blooms of phytoplankton growth from spring through to fall.

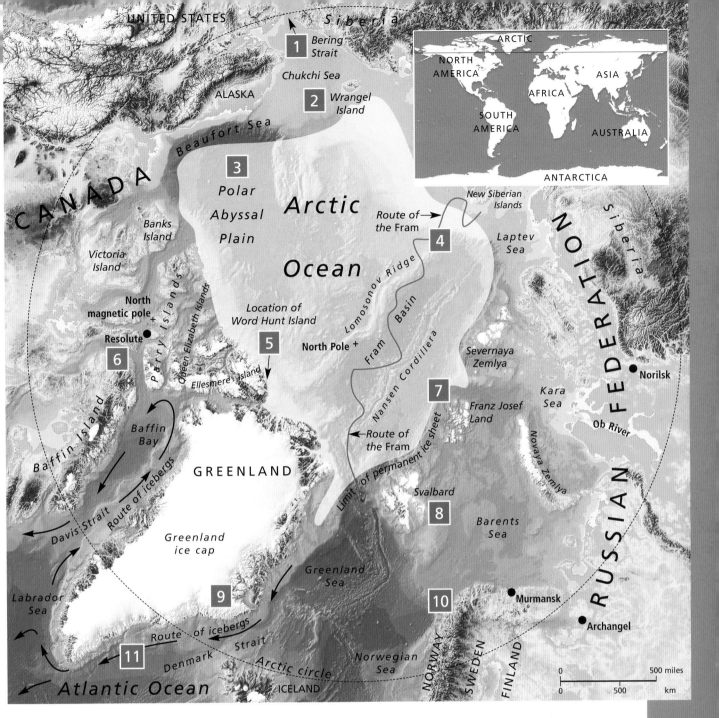

1. Bering Strait
A narrow passage between Alaska and far eastern Siberia.

2. Chukchi Sea
Walrus mothers and their calves swim from the Pacific through the Bering Strait in spring and follow the retreating ice north through the Chukchi Sea.

3. Polar Abyssal Plain
This deep plain lies 2.9 miles (4.6 km) beneath the arctic ice. That's 50 times the height of New York's Statue of Liberty.

4. Route of the *Fram*
The route of Norwegian explorer Fridtjof Nansen's ice-bound voyage in 1893–1896.

5. Word Hunt Island
Intrepid travelers often begin their North Pole expeditions here.

6. Resolute Bay
Beluga and narwhal whales pass by on their way to their summer feeding grounds in the Arctic. The small village of Resolute stands on the shore.

7. Permanent Ice Sheet
The ice sheet in the middle of the ocean breaks up, at its edges, into fragments called pack ice. In winter, the pack ice reaches Canada and Siberia.

8. Svalbard
People live on this remote group of islands. Some of the inhabitants still hunt whales.

9. Greenland Coast
Chilly arctic water flows close to this coast and keeps it ice-bound for most of the year.

10. Norwegian Coast
Atlantic water, warmed by the Gulf Stream and North Atlantic Drift, keeps the Norwegian coast free of ice although it is well within the arctic circle.

11. Greenland Icebergs
The International Ice Patrol tracks Greenland icebergs that might endanger shipping. The biggest icebergs reach about 1,500 feet (450 m) high and nearly 1 mile (1.6 km) long, with two-thirds or more lying unseen beneath the surface.

The Future

Everyone has a role to play in the fate of the oceans. What food we eat, how we get rid of garbage, and which politicians we vote for all have an impact on the health of our seas.

The 1990s were the warmest decade on record. Many scientists believe the global climate is getting warmer. If so, the likely cause is an enhanced greenhouse effect.

The greenhouse effect is a natural process that has been happening for billions of years. It is caused by gases in the atmosphere called greenhouse gases. When sunlight strikes Earth's surface, much of the energy is absorbed and then emitted back into space as heat. Some of the outgoing heat gets trapped in the atmosphere

Low-lying coral islands, including entire nations such as Tuvalu, could drown if sea levels rose as a result of excessive global warming.

by the greenhouse gases, warming the planet. The most important greenhouse gases are water vapor, carbon dioxide, and methane.

In the last 150 years, pollution has raised the level of some greenhouse gases. Extra carbon dioxide released by the burning of fossil fuels (coal, oil, and natural gas) has raised carbon dioxide levels in the atmosphere by more than a quarter. This has made the greenhouse effect stronger, possibly triggering global warming.

Rising Seas

Earth's climate goes through cycles over thousands of years. Sometimes factors combine to make the global climate cooler. When this happens, snow falls instead of rain and stays on the land instead of running into the sea. Ice builds up at the poles, and water in the oceans shrinks slightly as it gets cooler. As a result, sea levels fall. Scientists describe such times as ice ages, or glacials.

At other times, the global climate warms. When this happens, seawater expands, polar ice melts, and sea levels rise. We are in the

middle of one of these warm periods, called an interglacial. Since the peak of the last ice age, some 18,000 years ago, global sea levels have risen by 394 feet (120 m)—that's more than the height of the Statue of Liberty.

According to scientists' predictions, global warming could make Earth's climate some 4.5°F (2.5°C) warmer during the 21st century. The rise in temperature would make Earth's seawater expand slightly, and more polar ice could melt, so sea levels would rise. Many small tropical islands, such as the Marshall Islands in the South Pacific and the Maldives in the Indian Ocean, lie only 3–6 feet (1–2 m) above sea level. Such islands could disappear underwater within the next few centuries. Large

areas of Bangladesh, bordering the Bay of Bengal, are less than 6 feet (2 m) above sea level. Here, floods from storm waves already endanger many millions of people.

Poisoning the Seas

Pollution from land often ends up in the sea. Some pollutants are dumped there, others run into the sea through rivers, and air pollutants fall into the sea with rain.

Oil pollution is most obvious when a tanker runs aground and spills its cargo into the sea. But oil pollution is happening all the time at a low but harmful level: The amount of oil running into the sea from land is nearly ten times that from major tanker spills.

Some forms of pollution are particularly dangerous because they build up in the bodies of marine animals. This happens especially with dissolved heavy metals

? Global Meltdown

Some scientists are worried that global warming might be causing polar ice to melt faster. Most of Earth's ice is on land, on Greenland and Antarctica. If it melted more quickly, more water would run into the sea and sea levels around the globe would rise. For example, if the Lesser Antartica ice sheet were to collapse, sea

levels would go up by about 17 feet (5 m). Low-lying cities such as Miami would disappear under the sea. This scenario is unlikely in the near future, but there are many uncertainties about the pace of global warming. If the larger Greater Antarctica ice sheet ever collapsed, sea levels would rise by at least 170 feet (52 m).

Sea level 18,000 years ago, 394 feet (120 m) lower than today's.

Possible future sea level, 17 feet (5 m) higher than today's.

Possible future sea level, 170 feet (52 m) higher than today's.

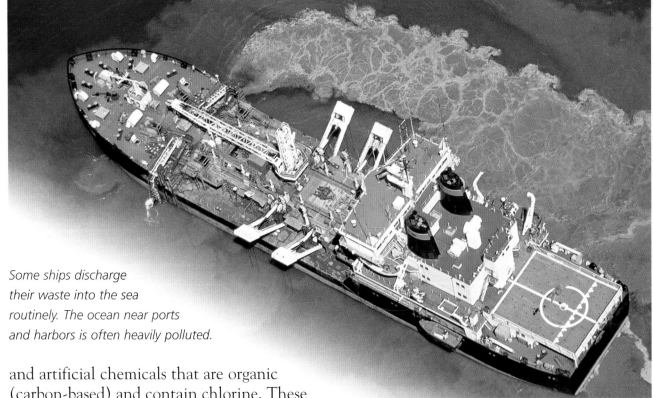

Some ships discharge their waste into the sea routinely. The ocean near ports and harbors is often heavily polluted.

and artificial chemicals that are organic (carbon-based) and contain chlorine. These organic chemicals include by-products from plastics manufacture, such as PCBs, and pesticides, such as DDT. PCBs might be responsible for the mysterious mass deaths that sometimes affect seals, sea lions, and dolphins. In 1987–1988, for instance, about 750 dead dolphins washed up on the east coast of the United States. Scientists found unusually high levels of PCBs in their bodies.

Global warming and pollution are not the only threats facing the oceans. Another is habitat loss—the destruction of fragile marine habitats, such as sea-grass meadows, coral reefs, and kelp forests. The worst habitat destruction occurs near shores or in shallow water, and the damage is often hidden from view. Trawling for fish and shellfish can strip the seafloor of almost all larger forms of life. It disrupts the lives of burrowing worms and shellfish, and it stirs up mud, smothering many small, bottom-living creatures.

Coral Grief
Coral reefs are very sensitive to human activities. Pollution and muddy water running into the sea from towns can kill coral. Tourists and fishers can damage coral by simply touching it, by standing on it, or by

dragging anchors or cables across it. Many reefs are overfished. Sometimes traders stun exotic fish with cyanide to gather them for the aquarium trade; the cyanide kills the coral.

In recent years, coral reefs have suffered from a mysterious condition called bleaching. The

This jackass penguin is a victim of an oil spill. It will probably fatally poison itself while trying to clean oil from its feathers.

Mass fish deaths can be a sign of a major pollution incident. Fish may be killed by toxic chemicals that leak into rivers and reach the ocean at estuaries.

coral organisms lose the microscopic algae that live inside them, which makes the coral turn a stark white, as though bleached. Death often follows. Scientists are not sure what causes coral bleaching, but many suspect a rise in sea temperature, caused by global warming, is to blame.

The Good News

The future for the oceans is not all bad news. Since the 1960s, governments across the world have set up more than 1,300 protected marine areas, though many of these are small and are not particularly well protected. Since 1994, most governments have also agreed to obey rules drawn up by the United Nations to protect the oceans. Meanwhile, scientific organizations such as the Intergovernmental Panel on Climate Change continue to monitor the oceans and publish findings, and many nongovernmental organizations, such as the World Wildlife Fund, raise awareness about marine pollution issues, overfishing, overhunting, and habitat loss.

 Lack of Action

At the 1992 Earth Summit in Rio de Janeiro, Brazil, many governments signed an agreement called the Convention on Climate Change, with the aim to cut the release of greenhouse gases. Progress since then has been slow. Industrialized countries such as the United States, Australia, Japan, and those in Europe produce far more greenhouse gases than countries in the developing world. The United States, for example, continues to produce about one-quarter of all industrial greenhouse gases.

Above: In Malaysia, as in many other countries, people deliberately remove coral for souvenirs or blow it up with dynamite for use as building material.

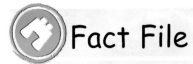 Fact File

▲ In 1997, an international survey called Reef Check revealed that nine-tenths of the world's coral reefs showed signs of damage from human activities.

▲ Worldwide, fishers lose or discard more than 100,000 tons of netting and line each year. Many turtles, seabirds, and marine mammals become entangled in these nets and die.

Energy from the Sea

The oceans contain many resources that could prove useful in the future. One is energy. The heat stored in the oceans, and the constant movement of seawater, are vast sources of energy that have been barely tapped. The technology to harness tidal power already exists. The Rance Estuary Barrage in France, for example, generates electricity by capturing the power of the tide. Each tide produces enough electricity for tens of thousands of homes.

Only a handful of tidal power stations have been built so far. At this one in France, the tide flows through the barrage at the far end, where the fast tidal current churns the water into white peaks. As the water passes, it drives turbines, which generate electricity.

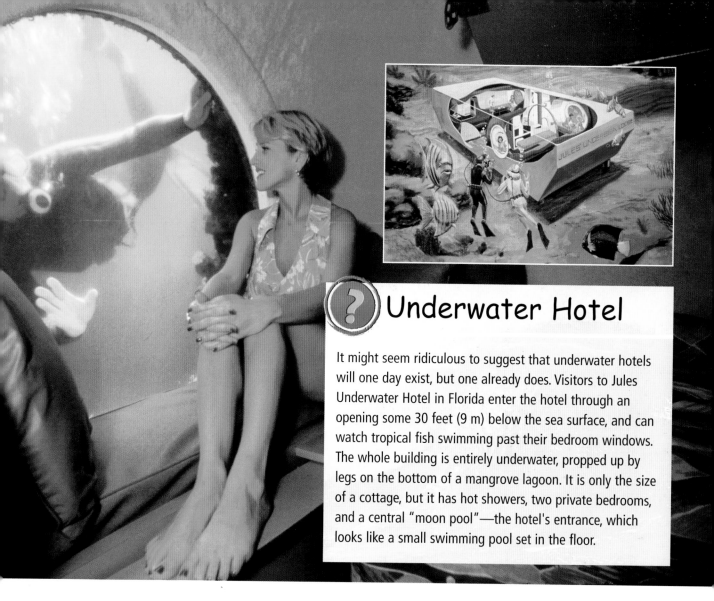

Underwater Hotel

It might seem ridiculous to suggest that underwater hotels will one day exist, but one already does. Visitors to Jules Underwater Hotel in Florida enter the hotel through an opening some 30 feet (9 m) below the sea surface, and can watch tropical fish swimming past their bedroom windows. The whole building is entirely underwater, propped up by legs on the bottom of a mangrove lagoon. It is only the size of a cottage, but it has hot showers, two private bedrooms, and a central "moon pool"—the hotel's entrance, which looks like a small swimming pool set in the floor.

Ocean energy sources cause less pollution than power stations driven by fossil fuels, but there are drawbacks. Wave-power stations are expensive to build, for instance, because they need to be strong enough to withstand storms but light enough to work with small waves. And tidal power stations can damage coastal habitats by altering the way the tide covers and uncovers beaches and mud flats.

New Minerals

Scattered over large areas of the deep-ocean floor are potato-sized lumps of valuable metals, such as cobalt, manganese, copper, and nickel. These lumps, or nodules, form over millions of years as dissolved metals slowly crystallize out of seawater. The nodules are worth trillions of dollars, but they are too expensive to harvest because they lie at least 2.5 miles (4 km) underwater. One day, when the demand for these metals is great enough, and mining technology is more advanced, the nodules may be scooped from the seafloor.

There are also metal deposits at vents in the seabed, sometimes in shallower water. Old vents that have stopped spurting hot water (and that no longer provide homes for the remarkable vent-living animals) could one day be mined. They are rich in zinc, copper, and even silver and gold.

The oceans are not just a pleasant place to visit—they are vital for the health of our planet. They provide an indispensable part of our food supply, but more than this, the circulating oceans, and the organisms within them, create our weather and our climate.

Glossary

alga (plural algae): A simple, plantlike organism that makes food from sunlight like a plant but lacks proper roots and leaves.

antibiotic: A substance, often a drug, that kills bacteria.

atmosphere: The layer of air around Earth.

atoll: A ring-shaped coral reef formed above a submerged volcanic island.

bacterium (plural bacteria): A single-celled microorganism; among the tiniest and simplest forms of life.

basin: A bowl-shaped depression on land or the seabed.

bioluminescence: Light created by living organisms.

biome: A major division of the living world, distinguished by its climate and wildlife. Tundra, desert, and temperate grasslands are examples of biomes.

carbon dioxide: A gas released when fuel burns. Carbon dioxide is one of the main gases thought to cause global warming.

cholesterol: A waxy chemical in the human body. When too much is present, it can block arteries.

climate: The pattern of weather that happens in one place during an average year.

community: A collection of organisms living in the same place, such as a sandy beach or a reef.

copepod: A tiny shrimplike animal that forms a major part of the animal plankton of the oceans.

delta: A wide, typically Δ-shaped plain at the mouth of a river where sediment collects.

downwelling: The downward movement of surface water to the depths of the ocean.

ecological: To do with the way organisms interact with one another and the environment.

equator: An imaginary line around Earth, midway between the North and South poles.

erode: To gradually wear away land by the action of wind, rain, rivers, ice, or the sea.

estuary: A place where the tide mixes with freshwater from a river, usually in an inlet of the sea at the mouth of a river.

evaporate: To turn into gas. When water evaporates, it becomes water vapor, an invisible part of the air.

gills: Organs that enable animals to breathe underwater.

global warming: The gradual warming of Earth's climate, thought to be caused by pollution of the atmosphere.

ice age: A period when Earth's climate was cooler and the polar ice caps expanded. The last ice age ended 10,000 years ago.

lagoon: An area of shallow water sheltered by a stretch of land.

larva (plural larvae): A young form of an animal. A tadpole is the larva of a frog.

larvacean: A tadpole-like marine animal related to sea squirts.

mangrove: A tree that grows in swamps on tropical coasts. Many mangroves have prop roots to support themselves.

methane: A gas that occurs naturally under the seabed. Also called natural gas, people sometimes use it as a fuel.

migration: A long-distance journey by an animal to find a new home. Many animals migrate each year.

nutrient: Any chemical that nourishes plants or animals, helping them grow. Marine plants absorb nutrients from seawater. Animals get nutrients from food.

oxygen: A gas in the air. Animals and plants need to take in oxygen so that their cells can release energy from food.

photosynthesis: The chemical process that plants and algae use to make food from simple chemicals and the sun's energy.

phytoplankton: Plantlike members of the plankton, such as microscopic algae.

plankton: Organisms that drift along with ocean currents, mainly near the surface.

polyp: A coral animal. Polyps are usually tiny but live together in large colonies of various shapes.

predator: An animal that catches and eats other animals.

protein: One of the major food groups. It is used for building and repairing plant and animal bodies.

rain forest: A lush forest that receives frequent heavy rainfall.

salp: A gel-like, jet-propelled sea animal related to sea squirts.

sea cucumber: A cigar-shaped relative of starfish. Many sea cucumbers resemble dill pickles.

sea grass: A true flowering plant that grows in meadows on seabeds just below low tidemark.

seamount: A submarine mountain rising from the seabed. Even its peak is underwater.

seep: A place on the seabed where chemicals such as oil and methane slowly enter the ocean.

spawn: To release lots of small eggs. Many fish and coral spawn.

species: A particular type of organism. Cheetahs are a species, but birds are not, because there are lots of different bird species.

symbiosis: An intimate partnership between two living organisms. Both partners may benefit, or only one may benefit.

temperate: Having a moderate climate. Earth's temperate zone lies between the warm tropical regions and the cold, polar regions.

tidal range: The difference in height between high and low tide.

trench: A canyon in the seabed.

tropical: Within about 1,600 miles (2,575 km) of the equator. Tropical places are warm all year.

tropical forest: Forest in Earth's tropical zone, such as tropical rain forest or monsoon forest.

tropical grassland: A tropical biome in which grass is the main form of plant life.

tundra: A biome of the far north, made up of treeless plains covered with small plants.

upwelling: The upward movement of water from the depths of the ocean to the surface.

vent: A hole in the seabed where high-pressure hot water and minerals enter the ocean.

zooplankton: Animal-like members of the plankton.

Further Research

Books

Byatt, A., Fothergill, A., and Holmes, M. *The Blue Planet: Seas of Life*. New York: DK Publishing, 2001.
Cousteau, Jacques Yves. *The Ocean World*. New York: Abradale Press, 1985.
Earle, Sylvia A. and Lindstrom, Eric. *Atlas of the Ocean: The Deep Frontier*. Washington, D.C.: National Geographic Society, 2001.

Websites

United States National Marine Sanctuaries: www.sanctuaries.nos.noaa.gov/oms/oms.html
(Features a clickable map and profiles of many beautiful sites in the ocean protected by the U.S. government.)
NOAA Ocean Explorer: oceanexplorer.noaa.gov/
(Follow the National Oceanographic and Atmospheric Administration as they explore the deep ocean.)
Sea and Sky: www.seasky.org/sea.html
(Packed with pictures, sounds, and information about the ocean, this is a colorful, fun, multipurpose site.)
Nova Online—Into the Abyss: www.pbs.org/wgbh/nova/abyss/
(Another fun exploration of the deep ocean.)
The Blue Planet Challenge—A Natural History of the Oceans: www.bbc.co.uk/nature/blueplanet/
(A huge resource packed with games and information about the oceans and their animals and plants.)

Index

Page numbers in *italics* refer to picture captions.

Picture Credits